Divided Solids Transport

There are no such things as applied sciences,
only applications of science.
Louis Pasteur (11 September 1871)

Dedicated to my wife, Anne, without whose unwavering support, none of this
would have been possible.

Industrial Equipment for Chemical Engineering Set

coordinated by
Jean-Paul Duroudier

Divided Solids Transport

Jean-Paul Duroudier

ELSEVIER

First published 2016 in Great Britain and the United States by ISTE Press Ltd and Elsevier Ltd

ISTE Press Ltd
27-37 St George's Road
London SW19 4EU
UK

www.iste.co.uk

Elsevier Ltd
The Boulevard, Langford Lane
Kidlington, Oxford, OX5 1GB
UK

www.elsevier.com

Notices

For information on all our publications visit our website at http://store.elsevier.com/

British Library Cataloguing-in-Publication Data
A CIP record for this book is available from the British Library
Library of Congress Cataloging in Publication Data
A catalog record for this book is available from the Library of Congress
ISBN 978-1-78548-183-3

Printed and bound in the UK and US

Contents

Preface

The observation is often made that, in creating a chemical installation, the time spent on the recipient where the reaction takes place (the reactor) accounts for no more than 5% of the total time spent on the project. This series of books deals with the remaining 95% (with the exception of oil-fired furnaces).

It is conceivable that humans will never understand all the truths of the world. What is certain, though, is that we can and indeed must understand what we and other humans have done and created, and, in particular, the tools we have designed.

Even two thousand years ago, the saying existed: "faber fit fabricando", which, loosely translated, means: "*c'est en forgeant que l'on devient forgeron*" (a popular French adage: *one becomes a smith by smithing*), or, still more freely translated into English, "practice makes perfect". The "artisan" (faber) of the 21st Century is really the engineer who devises or describes models of thought. It is precisely that which this series of books investigates, the author having long combined industrial practice and reflection about world research.

Scientific and technical research in the 20th century was characterized by a veritable explosion of results. Undeniably, some of the techniques discussed herein date back a very long way (for instance, the mixture of water and ethanol has been being distilled for over a millennium). Today, though, computers are needed to simulate the operation of the atmospheric distillation column of an oil refinery. The laws used may be simple statistical correlations but, sometimes, simple reasoning is enough to account for a phenomenon.

Since our very beginnings on this planet, humans have had to deal with the four primordial "elements" as they were known in the ancient world: earth, water, air and fire (and a fifth: aether). Today, we speak of gases, liquids, minerals and vegetables, and finally energy.

The unit operation expressing the behavior of matter are described in thirteen volumes.

It would be pointless, as popular wisdom has it, to try to "reinvent the wheel" – i.e. go through prior results. Indeed, we well know that all human reflection is based on memory, and it has been said for centuries that every generation is standing on the shoulders of the previous one.

Therefore, exploiting numerous references taken from all over the world, this series of books describes the operation, the advantages, the drawbacks and, especially, the choices needing to be made for the various pieces of equipment used in tens of elementary operations in industry. It presents simple calculations but also sophisticated logics which will help businesses avoid lengthy and costly testing and trial-and-error.

Herein, readers will find the methods needed for the understanding the machinery, even if, sometimes, we must not shy away from complicated calculations. Fortunately, engineers are trained in computer science, and highly-accurate machines are available on the market, which enables the operator or designer to, themselves, build the programs they need. Indeed, we have to be careful in using commercial programs with obscure internal logic which are not necessarily well suited to the problem at hand.

The copies of all the publications used in this book were provided by the *Institut National d'Information Scientifique et Technique* at Vandœuvre-lès-Nancy.

The books published in France can be consulted at the *Bibliothèque Nationale de France*; those from elsewhere are available at the British Library in London.

In the in-chapter bibliographies, the name of the author is specified so as to give each researcher his/her due. By consulting these works, readers may gain more in-depth knowledge about each subject if he/she so desires. In a reflection of today's multilingual world, the references to which this series points are in German, French and English.

The problems of optimization of costs have not been touched upon. However, when armed with a good knowledge of the devices' operating parameters, there is no problem with using the method of steepest descent so as to minimize the sum of the investment and operating expenditure.

Re-entrainment of Solid Particle Beds Swept by a Fluid Current

1.1. Introduction

1.1.1. *Significance of the problem*

The entrainment of sediments by water courses is a significant phenomenon in geography and, more generally, in earth sciences.

Re-entrainment affects various systems used in chemical and food industries, such as:

– gas cyclones and hydrocyclones;

– electrostatic precipitators, often found in cement works;

– decanter centrifuges equipped with disks (fast separators manufactured by Alfa Laval) if these devices are not fitted with radial separation bars (which is a mistake);

– convective microfiltration separating particles several micrometers in size using a membrane that is swept by pressurized suspension (nanofiltration is addressed by the laws of diffusion);

– filtration through a thick bed, which concerns the filtration of liquid suspensions. When the flow is very high, particles that have settled on the sand or coal grains are re-entrained.

1.1.2. *Method*

We can determine re-entrainment speed from the study conducted by Wu *et al.* [WU 03], adopting mean values for the geometric disposition of particles on the surface of the bed.

They also addressed the problem of saltation speed, that is the speed at which particles begin to move by a series of jumps, which we will not address here. However, we will present the major part of Wu's reasoning with regard to establishing the probability of re-entrainment, which is the fraction of the sediment bed surface whose particles are re-entrained.

Finally, we will consider the simple method proposed by Cao [CAO 97] for calculating the quantity of particles re-entrained per second and per unit of solid sediment bed surface, provided that the initial re-entrainment speed is known.

1.2. Shear velocity

1.2.1. *Piping*

The Reynolds number is given by:

$$Re = \frac{VD\rho}{\mu} = \frac{4W}{\mu P} \tag{1.1}$$

V: flow velocity (in empty vat) (m.s^{-1});

D: circular cross-section pipe diameter (m);

ρ: fluid density (kg.m^{-3});

μ: viscosity (Pa.s);

W: mass flow (kg.s^{-1});

P: wetted perimeter (m).

We verify equation [1.1] as:

$$\frac{V\rho\pi D^2}{4} = W \quad \text{and} \quad \pi D = P$$

The second expression of Re applies for the piping of any cross-section.

According to equation 6^e in Brun *et al.* [BRU 68, p. 57, Volume II], shear velocity is given by:

$$u_f = V\sqrt{\frac{\psi}{8}}$$

Turbulent flow with smooth shear is given by the Karman curve (p. 61):

$$\frac{1}{\sqrt{\psi}} = 2\log_{10}\left(Re\sqrt{\psi}\right) - 0.8$$

Semi-rough shear is given by the Colebrook formula (p. 69):

$$\frac{1}{\sqrt{\psi}} = 1.14 - 2\log_{10}\left(\frac{\varepsilon}{D} + \frac{9.32}{Re\sqrt{\psi}}\right)$$

Completely rough shear is given by:

$$\frac{1}{\sqrt{\psi}} = 2\log_{10}\frac{D}{\varepsilon} + 1.14$$

In theory, ψ must be calculated by each of the three methods, with the highest value being retained. In a turbulent regime, very often $\psi = 0.03$

NOTE.−

The pressure drop per unit of pipe length is given by (p. 52):

$$\frac{\Delta P}{L} = \frac{1}{D}\frac{\rho V^2}{2}\psi$$

1.2.2. *Flat plate (edge perpendicular to fluid flow)*

Here, the flow velocity reaches the velocity V_∞ away from the plate. Therefore, the Reynolds number is:

$$Re = \frac{V_\infty L \rho}{\mu}$$

L: the plate length along the flow.

The mean shear velocity along the plate is:

$$u_{fm} = V_\infty \sqrt{\frac{C_{fm}}{2}}$$

As with pipes, we may distinguish:

– laminar shear;

– smooth turbulent shear;

– semi-rough turbulent shear;

– rough turbulent shear.

In this last instance (p. 140):

$$C_{fm} = 0.0081 \left(\frac{\varepsilon}{L} \right)^{1/7}$$

Smooth turbulent shear is given by:

$$C_{fm} = 0.074 Re^{-0.2} \quad \text{when } Re < 10^7$$

$$C_{fm} = 0.455 \left(\log_{10} Re \right)^{-2.58} \quad \text{when } Re > 10^7$$

In practice, C_{fm} is calculated by both of these methods, with the highest value being retained.

1.3. Strength and lift

1.3.1. *Hypotheses*

According to Wu *et al.* [WU 03], the base of the particles (which are assumed to be spherical) is situated at the level δ relative to the bed surface. Furthermore, these authors show that the level δ can vary between two extremes:

$-0.75d < \delta < 0.116d$, which is, on average, $0.31d$

where d is the diameter of the particle. The center of the particle is found, on average, at level:

$$y = \frac{d}{2} + \overline{\delta} = d(0.5 - 0.31) = 0.19d$$

The velocity profile is given universally according to the level:

$$u(y) = u_f \left[5.5 + 5.75 \, \log_{10} \left(\frac{u_f y \rho}{\mu} \right) \right]$$

1.3.2. *Mean thrust velocity*

The elementary surface (Figure 1.1) is:

$$dA = 2\frac{d}{2}\sin\psi dy = d \, \sin\psi d \left(\frac{d}{2}\cos\psi \right) = \frac{-d^2}{2}\sin^2\psi d\psi$$

Since $\delta < 0.116d$, we still have $\delta < 0.25d$, which limits the angle ψ to the angle ψ_0, such that:

$$-\frac{d}{2}\cos\psi_0 = \delta + \frac{d}{2} - 0.25d = \delta + 0.25d$$

$$\cos\psi_0 = -\frac{(2\delta + 0.5d)}{d}$$

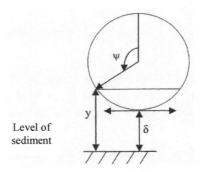

Figure 1.1. *Auxiliary angle ψ*

The exposed surface is the surface of a circle set by the chord subtended by the angle $2\psi_0$:

$$A = \frac{-d^2}{2} \int_0^{\psi_0} \sin^2\psi d\psi = \frac{-d^2}{2}\left(\frac{\psi_0}{2} - \frac{\sin 2\psi_0}{4}\right)$$

On the contrary:

$$\cos\psi = \frac{y - 0.5d - \delta}{0.5d} \text{ and } \sin^2\psi = 1 - \cos^2\psi = \frac{(d + \delta - y)(y - \delta)}{0.25d^2}$$

The mean thrust velocity is:

$$\bar{u} = \frac{1}{A} \int_A u dA,$$

with

$$\int_A u dA = \int_{0.25d}^{d+\delta} u(y) \frac{\sqrt{(d + \delta - y)(y - \delta)}}{5d} dy$$

This integral can only be calculated numerically.

Thus, we have:

$$\bar{u} = \bar{u}(u_f, d, \delta)$$

1.3.3. *Drag force and lift force*

The drag force is the force exerted by the fluid on a particle in the direction of fluid movement parallel to the sediment surface. It is given by:

$$F_T = C_X \frac{\rho \bar{u}^2}{2} A$$

Many expressions exist for C_X. Here, we will use that proposed by Schiller *et al.* [SCH 33]:

$$C_X = \frac{24}{Re_p} \left(1 + 0.15 Re_p^{0.687}\right)$$

with

$$Re_p = \frac{\bar{u} d_p \rho}{\mu}$$

Lift force tends to push the particles away from the divided solid bed in a perpendicular direction:

$$F_S = C_x \frac{\rho \bar{u}^2}{2} A$$

Wu *et al.* [WU 03] accept that the lift force and drag force are equal.

1.4. Entrainment criteria

1.4.1. *Attachment forces*

These forces tend to maintain the cohesion of the bed, that is they keep the particles in the bed. These forces are:

1) Gravity, which is the weight of the particle:

$$F_G = \frac{\pi d^3}{6} g (\rho_S - \rho)$$

If the bed is inclined horizontally, we must apply the gravity component perpendicular to the bed.

2) Van der Waals forces:

$$F_A = \frac{AD}{12h^2}$$

$$A = \left(A_1 A_2\right)^{1/2} \qquad \frac{1}{D} = \frac{1}{D_1} + \frac{1}{D_2}$$

A_1 and A_2: the Hamaker constants of two bodies in contact.

We take:

$$A = 2.10^{-18}\,J$$

D_1 and D_2: the diameter of two particles in contact.

Here, we will calculate F_A assuming that the particle is in contact with a flat surface.

h: the distance between the two surfaces.

We take:

$$h = 50\,nm.$$

3) Percolation

On its surface, the bed is crossed orthogonally by a liquid that crosses the membrane supporting the solid sediment.

In this situation, the particle's diameter is in the order of micrometers.

The empty vat velocity of the liquid crossing the bed is in the order of $1\ \mu m.s^{-1}$.

Accordingly, the regime is laminar and the force exerted on the particle is:

$$F_P = 3\pi d\mu\left(U/\varepsilon\right) \qquad \left(r = d/2\right)$$

V: empty vat velocity of liquid across the bed (m.s^{-1});

ε: sediment porosity (in the order of 0.4);

μ: liquid viscosity (Pa.s).

1.5. Quantity entrained

1.5.1. *Probability and frequency*

1.5.1.1. *Level δ*

We have seen that δ varies between −0.75d and +0.116d, i.e. a range of 0.866d.

The proportion of δ levels included between δ and d(δ) is assumed to be:

$$f(\delta)d(\delta) = \frac{d(\delta)}{0.866d}$$

where f(δ) is zero outside of the above interval.

1.5.1.2. *Mean velocity ū*

We accept that $\bar{u}$ is apportioned according to a log-normal law. Accordingly, we write:

$$v = \text{Ln}\, u \quad \bar{v} = \text{Ln}\, \bar{u}(d,\delta) \quad \sigma = \sqrt{0.128} = 0.3578$$

The probability that a particle (d, δ) will be entrained is given by:

$$P(u > u_e) \qquad\qquad = 1 - P(0 < u < u_e)$$

$$= 1 - P(-\infty < v < v_e)$$

$$= 1 - \int_{-\infty}^{v_e} \frac{1}{\sqrt{2\pi}\sigma} \exp\left[-\frac{(v - \bar{v})^2}{2\sigma^2}\right] dv$$

Therefore, according to Cheng *et al.* [CHE 98], we can write:

$$\frac{1}{\sqrt{2\pi}} \int_0^x \exp\left(-\frac{t^2}{2}\right) dt \# 0.5 \frac{x}{|x|} \sqrt{1 - \exp\left(-\frac{2x^2}{\pi}\right)}$$

Hence,

$$P = (u > u_e) = P(v > v_e) = 1 - 0.5 - 0.5\varepsilon \sqrt{1 - \exp\left(-\frac{2(v_e - \bar{v})^2}{\pi\sigma^2}\right)}$$

$$\varepsilon = 1 \text{ if } v_e > \bar{v}$$

$$\varepsilon = -1 \text{ if } v_e < \bar{v}$$

Of course,

$$P(v > v_e) = P(d, \delta)$$

1.5.2. *Results of Wu et al.*

Wu *et al.* [WU 03] considered the probability of entrainment (see Figure 1.2) due to the dispersion of turbulent fluid velocities according to a reduced shear stress:

$$\theta = u^2_f / u^2_{f0},$$

where u_{f0} is the shear velocity at which $P = 1$.

Recall that shear stress on the wall is:

$$\tau_p = \rho u_f^2,$$

ρ: fluid density $(kg.m^{-3})$;

u_f: shear velocity.

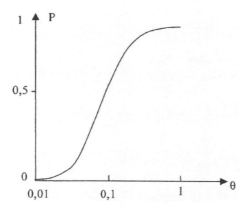

Figure 1.2. *Probability of entrainment*

1.6. Entrainment limit velocity (simplified method)

1.6.1. *Mean geometry of spherical particles on the surface of the bed*

The surface of the particle bed is irregular and, according to its mean level (Figure 1.3), the mean value for the lowest point of a particle is:

$$\frac{\delta}{d} = \frac{1}{2}(-0.75 + 0.116) = -0.317$$

The mean value for the center of gravity of the exposed surface is taken as being equal to 0.28d. The covered part is not subjected to the action of the fluid flow.

The level of the center is:

$$d(-0.317 + 0.5) = 0.183d$$

The exposed surface is:

$$\frac{\pi d^2}{4} - \frac{d^2}{8}(\theta - \sin\theta)$$

with

$$\cos\theta = \frac{0.183}{0.5} = 0.366 \quad \text{hence} \quad \theta = 1.196 \text{ rad}$$

and

$$\sin\theta = 0.93$$

The exposed surface is:

$$A_{ex} = d^2 \left(\frac{\pi}{4} - \frac{(1.196 - 0.93)}{8} \right) = 0.7529 \text{ d}^2 \qquad [1.2]$$

$$A_{ex} = k_{ex}d^2 \quad \text{with} \quad k_{ex} = 0.7529$$

The equivalent hydraulic diameter is:

$$D_{eq} = \frac{4A}{P} = \frac{4 \times 0.5724}{2.876} = 0.8d$$

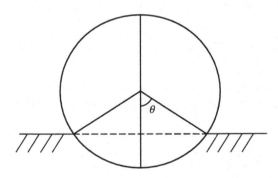

Figure 1.3. *Mean position of particle*

1.6.2. *Velocity of sweeping for entrainment*

F_F is the total of the attachment forces (see section 1.4.1):

$$F_T = F_G + F_A + F_P$$

We write that the lift force F_S necessary for entrainment is equal to the drag force F_T:

$$F_S = F_T = \frac{\rho u^2}{2} C_X A_{ex}$$

Entrainment then happens if:

$$F_S \geq F_T$$

which is to say:

$$\frac{\rho u^2}{2} C_X A_{ex} \geq F_T \qquad\qquad [1.2]$$

According to section 1.3.3:

$$\frac{1}{C_X} = \frac{Re_p}{24(1+\alpha)} \quad \text{with} \quad \alpha = 0.15\, R_{ep}^{0.687}$$

The Reynolds number for the particles is:

$$Re_p = \frac{u\, d\, \rho}{\mu}$$

therefore:

$$\frac{1}{C_x} = \frac{u\, d\, \rho}{24\mu(1+\alpha)}$$

and:

$$A_{ex} = k_{ex} d^2$$

Equation [1.2] then becomes:

$$\frac{\rho u^2 \times 24 \times (1+\alpha)}{2 \times U\, d\, \rho} \mu k_{ex} d^2 = F_T$$

hence:

$$u = \frac{F_T}{12(1+\alpha) k_{ex} \mu d} \qquad\qquad [1.3]$$

The velocity u is taken from the center of the particle, which is level y = 0.19d (see section 1.3.1).

The friction velocity u_f is defined by:

$$u = u_f \left[5.5 + 5.75 \log_{10} \left(\frac{u_f y \rho}{\mu} \right) \right] \quad \text{with} \ \ y = 0.19d \ \left(\text{see section 1.3.1} \right)$$

(see [BRU 68, p. 133].)

The friction coefficient ψ is defined in section 1.2.1.

The entrainment velocity V_E is:

$$V_E = u_f \sqrt{\frac{8}{4}}$$

The coefficient ψ is often in the order of 0.03.

EXAMPLE I.–

The fluid is the air flowing through a duct with:

$d = 10^{-4}\,\text{m}$ $\qquad \mu = 20.10^{-6}\,\text{Pa.s}$ $\qquad \rho = 1.3\,\text{kg.m}^{-3}$

$D = 0.0635\,\text{m}$ $\qquad \rho_s = 2,700\,\text{kg.m}^{-3}$

Calculation of the attachment forces:

1) Weight:

$$F_G = \frac{\pi}{6} \times 10^{-12} \times 2,700 \times 9.81 = 1.387 \times 10^{-8}\,N$$

2) Van der Waals attraction:

The Hamaker constant is taken as 2×10^{-18} J and the distance h to the surfaces as 50 nm:

$$F_A = \frac{Ad}{12\,h^2} = \frac{2 \times 10^{-18} \times 10^{-4}}{12 \times 25.10^{-16}} = 0.666.10^{-8}\,N$$

The total of the attachment forces (see section 1.4.1) is:

$$F_F = (1.387 + 0.666)10^{-8} = 2.053 \times 10^{-8}\,N$$

$$u^{(o)} = \frac{2.053 \times 10^{-8}}{12(1+\alpha) \times 0.7529 \times 10^{-4} \times 20.10^{-6}} = \frac{1.13\ m.s^{-1}}{1+\alpha}$$

$$Re_p^{(o)} = \frac{1.13 \times 0.0635 \times 1.3}{(1+\alpha) \times 20.10^{-6}} = \frac{4664}{1+\alpha}$$

$$\alpha = 0.15\left(\frac{4664}{1+\alpha}\right)^{0.684}$$

Tests show that $\alpha \gg 1$:

$$\alpha^{1.684} = 0.15 \times 323.16 \quad \alpha = 10.03$$

so:

$$u = \frac{1.13}{11.03} = 0.102\ m.s^{-1}$$

The shear velocity u_f is as it is at the center of the particle :

$$0.102 = u_f \left[5.75 \log_{10} \left(\frac{u_f \times 0.19 \times 10^{-4} \times 1.3}{20.10^{-6}} \right) + 5.5 \right]$$

The solution of this equation is:

$$u_f = 0.125 \text{ m.s}^{-1.}$$

hence:

$$V_E = 0.125 \sqrt{\frac{8}{0.03}} = 2.04 \text{ m.s}^{-1.}$$

EXAMPLE II.–

The fluid is water flowing through piping with:

$$d = 10^{-3} \text{ m} \qquad \mu = 10^{-3} \text{ Pa.s} \qquad \rho = 1,000 \text{ kg.m}^{-3}$$

$$D = 0.032 \text{ m} \qquad \rho_{eff} = 2,700 - 1,000 = 2,700 \text{ kg.m}^{-3}$$

$$F_G = 13.9.10^{-6} \text{ N}$$

$$F_A = 0.066.10^{-6} \text{ N}$$

$$F_T = (13.9 + 0.066)10^{-6} = 13.966.10^{-6} \text{ N}$$

$$u = \frac{13.966^{10^{-6}}}{12 \times (1+\alpha) \times 0.7529 \times 10^{-3} \times 10^{-3}} = \frac{1.54}{1+\alpha}$$

$$Re = \frac{1.54 \times 0.032 \times 1000}{(1+\alpha)10^{-3}} = \frac{0.0493.10^6}{1+\alpha}$$

$$\alpha = 0.15 \left(\frac{0.0493.10^6}{1+\alpha} \right)^{0.687}$$

$$\alpha \gg 1 \qquad \alpha^{1+0.687} = 251 \qquad \alpha = 251^{0.592} = 26$$

so:

$$u = \frac{1.54}{26+1} = 0.057 \text{ m.s}^{-1}$$

The shear velocity at the center of the particle is u_f, given by:

$$0.057 = u_f \left[5.75 \log_{10} \left(\frac{u_f 0.19 \, d\rho}{\mu} \right) + 5.5 \right]$$

The solution to this equation is:

$$u_f = 0.0085 \text{ m.s}^{-1}$$

Hence:

$$V_E = 0.0085 \sqrt{\frac{8}{0.03}} = 0.13 \text{ m.s}^{-1}$$

1.6.3. *Angular particles*

The above work is only concerned with spherical particles. According to Zenz [ZEN 64], the entrainment velocities of angular particles (wheat, ervil seed, rapeseed, clay pellets, sand, salt) are half those of smooth spheres. Indeed, the coefficient C_x is twice higher for irregular forms.

1.6.4. *Cleaning speed and non-entrainment velocity*

Cleaning speed is the velocity of a fluid that evacuates the sediment in its entirety. We suggest:

$$V_{clean} = 3V_e$$

The non-entrainment velocity is the velocity below which the sediment remains intact. We suggest:

$$V_{ne} = V_e/2$$

The velocities of pneumatic transport and hydraulic transport are often in the orders of 10 and 1 $m.s^{-1}$ and, consequently, higher than V_{clean} as the corners need to be swept.

1.7. Entrained flowrate

1.7.1. *Probability of entrainment and entrained flow [CAO 97]*

The total number of molecules in 1 m^2 of bed surface is:

$$n_o = \frac{1-\varepsilon}{\left(\dfrac{\pi d^2}{4}\right)}$$

Among these, the proportion of mobile particles is P, which is known as the probability of entrainment. The volume of entrained particles in movement is then applied to the unit of bed surface:

$$Q_e = \frac{V^2}{100v} n_o P \frac{\pi d^3}{6} = \frac{2(1-\varepsilon)P d V^2}{300v} \qquad \left(m^3.s^{-1}.m^{-2}\right)$$

P: the numerical fraction of the surface particles subject to entrainment;

V: the empty vat velocity of the fluid $(m.s^{-1})$.

According to Cao [CAO 97]:

$$P = 0.02\left(V/V_e\right)^2$$

however, it is equally possible to apply the expression from Wu *et al.* [WU 03].

In order for P to remain ≤ 1, we must have:

$$V/V_e \leq 7.07$$

d: the particle diameter (m);

ε: the porosity of the particle bed (where the void fraction is the volume);

v: the fluid kinematic viscosity (m^2.s^{-1}), with

$$v = \mu_f / \rho_f$$

μ_f: the fluid dynamic viscosity (Pa.s);

ρ_f: the fluid density (kg.m^{-3});

V_e: entrainment start velocity (m.s^{-1}).

2

Mechanical Conveying of Divided Solids

2.1. General

2.1.1. *Separation due to fall and pouring: dust formation*

In general, disparities in pouring fall velocity and rolling velocity on a slope are a major cause of separation.

Particles of size <60 μm tend to remain suspended in the air during a fall, leading to the formation of dust on horizontal surfaces following device deactivation. This occurs with long processes and, in particular, if one of the components involved is friable, which can lead to the formation of fines by attrition. Products that give rise to dust are known as dust-making products.

Plinke *et al.* [PLI 94] studied dust formation. Operations that involve milling, mixing or sand-jet cleaning produce fines. The mass of dust created depends on the velocity hitting the target, humidity, size distribution and the nature of the divided solid. More specifically, it depends on the ratio:

$$\frac{\text{Separating force}}{\text{Binding force}}$$

The authors admit that the separating force is an increasing function of the arrival velocity, which is not necessarily equal to $\sqrt{2gh}$. They calculated this velocity using the force exerted on the balance pan at the instant of dropping.

The arrival velocity is a function of:

– drop height (h);

– flowrate;

– median diameter;

– true density;

– angle of repose;

– cone diameter on arrival.

2.1.2. *Flowrate and storage capacity combinations (notions)*

In addition to the fact that flowrate can be expressed in mass or apparent volume, we may distinguish:

– nominal or mean flow, corresponding to ideal operation;

– peak flow (for a discontinuous installation);

– operating or real flow.

The flowrate used to determine the equipment is the calculation flowrate that is either the peak flow or the nominal flow increased by 10–20%, according to the situation.

However, the next logistical problem is not easily resolved.

A production using mean flow W is supposed to be continuous 24 h a day. This production is shared between n destinations with mean flowrates W_i:

$$W = \sum_{i=1}^{n} W_i$$

Every day, each destination only accepts the share of production that it is due for activity time $t_{Ai} < 24$ h. Therefore, we need to have:

– the instantaneous flow W_{Ii} toward each destination so that:

$$W_{Ii} t_{Ai} = W_i \times 24 \qquad \left(W_{Ii} > W_i \right)$$

– the product to be stored, arriving during inactive time $t_{Ii} = 24 - t_{Ai}$ which represents mass:

$$M_i = W_i t_{Ii} = W_i \left(24 - t_{Ai} \right)$$

The design of the installation can be further complicated to account for stoppages, the possibility of breakdowns, several days of inactivity, declines in production, non-compliance with norms, etc.

Reasonings such as these help to determine whether flowrates and storage between a continuous receipt and discontinuous sources.

2.1.3. *Organization of divided solid circulation*

Krambrock [KRA 79] studied the following facets:

– capacity filling or charging installations;

– emptying installations (desilage);

– type and cost of investment for storage recipients together with silo or hopper level measuring;

– transporting the divided solid or recipient holding it;

– bagging the end products.

Johanson [JOH 78] reviewed the various systems for taking the product, whether it is:

– stored in heaps;

– silos with one or several udders (outlets);

– portable bins.

Thomson [THO 79] conducted a critical review of the various types of conveyors (which he called "feeders"). In particular:

– rotary vanes;

– screws;

– vibrating channels;

– conveyor belts.

We will briefly consider extractors positioned below storage capacities (hoppers or silos).

2.1.4. *Extractors*

Here, we will complete Johanson's study [JOH 69].

These devices, which act as conveyors for loose solids, are positioned beneath the extraction point of the storage capacity. A correctly designed extractor must be capable of both accepting the product and evacuating it without any clogging or blockage.

The extractor must take the product under the whole of the orifice cross-section, which minimizes preferential channels in the convergent (and thereby a possible cause of segregation).

The extractor's feed opening (which is referred to as the "mouth", similar to crushers) must be large enough to make the evacuated flow independent of the product's properties (its flowability in particular) and to withstand any arch collapse that may occur.

There must be a flowrate safety margin for this reason.

The choice of extractor type depends on the configuration of the mouth.

1) Beneath apertures, we often fit a conveyor belt running along the aperture. The vertical distance between the belt and the skirt edge must be larger than the size of the largest particles, but it must not be very high so as to avoid spilling from the belt edges. We can fit vertical flaps perpendicular to the direction of movement, arrayed in series and staggered with a vertical distance to the wider downstream belt. Thus, flow can increase in the direction of movement, and extraction can occur all along the aperture.

2) Belt extractors are to be avoided for fragments ("particles" larger than 0.15 m) as they cannot withstand the fall impact. In such cases, we use a metallic apron feeder conveyor, which is more expensive but can withstand both impacts and extreme temperatures.

3) Screws only "draw" the product under a square, horizontal surface whose side is around 1.5 times that of the screw diameter. As a consequence, the screw must preferably be associated with round or square mouths.

However, screws may be used beneath an orifice, provided their flowrate is able to increase downstream (increasing the screw pitch or diameter or alternatively, decreasing the shaft diameter).

4) Rotary vanes are only used under round or square holes. In such cases, they are inserted between the hole and the extractor, thereby allowing for regulation of the extracted flow (except in cases where the stored mass is transpierced by an axial well).

5) Extractors with paddles in the same direction as the (horizontal) rotation axis are suitable for pneumatic conveyance in a duct, whose axis and diameter are the same as those of the extractor. The conveyed flowrate can be regulated continuously and effectively.

6) Vibrating extractors, like belt extractors, require increasing vertical spacing upstream between the orifice and trough. If spacing is too small, the vertical component is disturbed, which blocks the system.

Sometimes, these devices should be avoided for fine products, which have a tendency to pack with the effect of vibration, thereby blocking the flow.

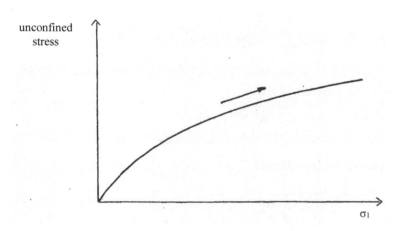

Figure 2.1. *Blockage due to overpressurization*

Overpressures caused by vibration increase the product's unconfined stress, f_c, see Figure 2.1. In order to flow, a higher value is needed for the trough slope. With fine products that it is best to reduce the intensity of environmental vibrations (machines).

7) Rotary tables with a plow are only suited to circular or square orifices and for relatively fine products, since the product must be able to flow between the lower edge of the vertical cylindrical skirt extending the orifice. The product is then evacuated by a plow share set outside of the skirt that scrapes the rotating table.

We will study the following forms of mechanical conveyance in detail:

– chutes and airslides;

– conveyor belt;

– chain conveyors;

– vibrating conveyors;

– lifts;

– transport in batches.

2.2. Chutes and airslides

2.2.1. *Chutes inclined horizontally*

The significant parameters for the flow of an inclined chute with a rectangular cross-section are:

– inclination angle α to the horizontal axis;

– the vertical opening h of the upstream feed orifice;

– chute length L and width ℓ;

– particle diameter d_p.

Campbell *et al.* [CAM 85] observed that the nature of flow changes depending on the opening h. Indeed, as h increases from zero, the flow of solid is limited while particle velocity V_p is significant, such that the product bed thickness in the flow is limited:

$$V_p = \sin\alpha\sqrt{gD_H} \qquad\qquad D_H = \frac{4A}{P} = \frac{4\ell h}{(\ell + 2h)}$$

D_H: the hydraulic diameter of upstream opening.

If h continues to increase, the solid flow as mass W increases up to a limit W_T, which depends on α, ℓ and d_p. Upstream conditions, i.e. the opening h, no longer intervenes. Velocity is limited and bed thickness is significant. For this last type of flow (that we will call the local regime, since neither upstream nor downstream conditions apply), the influence of chute incline α is as follows:

– If the slope is only slightly higher than the angle of repose, the regime is slipping not established. Particle velocity is low and we observe the percolation of fines and angulars at the bottom of the chute, where they form an immobile layer. A mobile layer slips on top of this layer. The thickness of the stationary layer decreases downstream, so that the real slope is higher than the chute's slope. The velocity profile is concave.

– As the slope increases, the regime is slipping established and the dead zone disappears. The velocity profile becomes linear.

– If the slope increases again and significantly exceeds the angle of repose, flow takes an agitated ("turbulent") aspect, and we can observe intense jumping of surface particles. Here, the velocity profile is convex with strong gradients both at the surface and beneath it.

NOTE.–

The transition described above corresponds, as we have seen, to a flowrate W_T. The W_T/ℓ ratio changes in the following ways:

– increases with ℓ;

– increases with α;

– decreases with d_p.

In Figure 2.2(a) and (b), we see that the single variation curve for W/ℓ is subdivided into a "harp" as h increases and into transition points such that W becomes W_T.

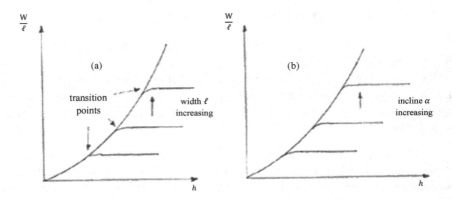

Figure 2.2. *Variation of W/ℓ with opening h:*
a) parameterization as ℓ; and b) parameterization as α

This transition occurs if the chute incline decreases at the bottom. The product slows, bed thickness increases abruptly and an immobile layer can develop at the bottom of the chute. However, we can avoid this by widening the chute exit, which reduces the thickness of the solid bed.

2.2.2. *Vertical chute*

Vertical chutes are often spherical with a diameter systematically greater than 0.15 m in order to avoid the formation of solid blockages.

Firewicz (according to [WEI 91]) proposed the following empirical relationship for volume flowrate:

$$Q = 10D^{2.79}$$

Q: low volume ($m^3.s^{-1}$);

D: chute diameter (m).

EXAMPLE.–

$$D = 0.15 \text{ m}$$

$$Q = 10 \times 0.15^{2.79} = 0.05 \text{ m}^3.s^{-1} = 50 \text{ L.s}^{-1}$$

2.2.3. *Interruption or modulation of chute flow*

These actions only apply to vertical sections of chutes.

2.2.3.1. *Interruption (sectioning)*

Slide valves with a guillotine-shaped shutters are widely used.

We can also fit a butterfly valve after the slide valve if a gas seal is required.

2.2.3.2. *Modulation*

Rotary vanes are widely used in order to control the extraction flowrate from a hopper for instance. These can be:

– fed from above with exit below; or

– fed from above with a horizontal exit.

In both designs, the rotor includes sectors that are fed from above, although in the first instance discharge occurs by means of gravity, while in the second, a gas impulse (usually air) expels the product while the position of the sector is low.

If the product is fine (d_p < 100 μm), or if it includes scales or plates, there will inevitably be a leak between the rotor and the stator, since the clearance provided varies between 75 and 250 μm. In such cases, we can edge the rotor vanes with an elastomer. In addition, we need to protect the bearings with a system of double seals with an intermediate lantern. A continuous insufflation of air can suffice.

If the product is hot and if the vane is cold, the clearance between the rotor and the stator can take an unexpected value.

If the product is sticky, we can fit shallower sectors for easier discharge, but it would be better when used with a horizontal exit, if possible.

If the product is cohesive, the chute should be oversized in order to avoid arching. In addition, shallow sectors should be used in order to keep the flowrate to a limited value.

2.2.4. *Airslides*

These are chutes that are slightly horizontally inclined (from 3 to 10 degrees) and typically square shaped (e.g. 0.3 × 0.3 m). However, the width of airslides can vary from 0.15 to 1 m.

These conveyors are seldom longer than 30 m. Essentially, air is injected along their lower side, which is porous for this purpose. It aerates the product, thereby making it fluid and facilitating its slide under the influence of gravity. An extremely long conveyor would lead to excessive longitudinal velocity of the injected air, which would accumulate in the airslide. The flow of injected air is:

$$Q = V_o L \ell,$$

Q: flow in the airslide under internal conditions (m³.s⁻¹);

L: airslide length (m);

ℓ: airslide width (m);

V_o: injection velocity (m.s⁻¹), with

$$0.0025 \text{ m.s}^{-1} < V_o < 0.008 \text{ m.s}^{-1}$$

These velocities correspond to a drop in pressure of several kilopascals on crossing the porous lining.

These conveyors, whose price and maintenance costs are low, are recommended for fine and relatively fine products:

$$20 \text{ μm} < d_p < 500 \text{ μm}$$

Ultrafines ($d_p < 20$ μm) fluidize poorly, while grains ($d_p > 500$ μm) would not be lifted by the gas current.

EXAMPLE.–

$$\ell = 0.3 \text{ m} \quad L = 30 \text{ m} \quad V_o = 0.003 \text{ m.s}^{-1}$$

$$Q = 0.003 \times 0.3 \times 30 = 0.027 \text{ m}^3. \text{ s}^{-1}$$

Air velocity on airslide exit is given by:

$$V_{GS} = \frac{0.027}{(0.3)^2} = 0.3 \text{ m s}^{-1}$$

2.3. Conveyor belts

2.3.1. *Description*

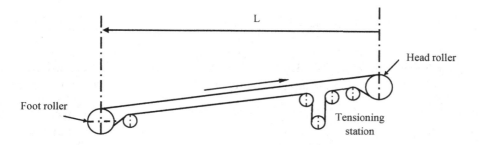

Figure 2.3. *Conveyor belt*

A conveyor belt is depicted in Figure 2.3. It is important to note that the guiding roller, which exerts the motor power, is typically the head roller; however, it is not always necessary.

Both the driving side (top) and return side are held in place by support rollers (not shown in the figure). Their spacing varies from 0.7 to 1.7 m. This decreases with the increase of belt width and the apparent density of the solid.

Regulation of belt position is significant. The belt must not veer laterally. Stability is ensured by shifting any slack to the rollers. The belt position is fixed by the rollers supporting the return side. These rollers can be orientated along any lateral direction to which we wish to shift the belt. The pinch used is very low, being 2°–3°.

The belt is made up of a certain number of successive layers known as folds. These folds can be made of cotton, polyamide or polyester.

These layers are stuck together by gumming and can provide mechanical resistance to the belt. Together, the layers are coated on both sides with protective elastomer leaf, according to temperature, composed of:

– chloroprene (up to 135°);

– butyl (up to 170°);

– silicone rubber (up to 250°).

Certain coatings are able to withstand hydrocarbons (oils, fats, etc.) or abrasive products. For food products, belts must neither produce any toxic material nor confer any taste or smell (i.e. no flavor) to the transported product.

The thickness of the top coating increases with particle size and also with the number of turns made by the belt per unit of time.

For rubber, this thickness varies between 1.5 and 6 mm. Thicknesses can be divided by 3 for polyvinyl chloride, since it presents superior resistance to abrasion.

The belt's mass per m^2 depends on the number and nature of folds, and the thickness and nature of the coating. Cotton folds are the heaviest. Mass per m^2 varies from approximately 5 m.m^{-2} to 20 kg.m^{-2}.

2.3.2. *Use criteria*

The advantages of conveyor belts are:

– they allow for capture beneath long apertures (of up to 5 m);

– they can, though not continuously, withstand the impact of larger blocks.

On the contrary:

– their path must be straight;

– the rising slope is limited to 30°;

– the elastomer range limits product temperature.

2.3.3. *Flow regulation by belt scale*

A belt scale can be fitted beneath the roller system in order to measure the mass of the product present on a given length of belt. The belt's velocity is measured by a tachometer. The result of these two measurements provides a mass flowrate precision of ±1%.

A calculation can also be made of total flow over time.

2.3.4. *Safeguards*

1) A lateral pathway parallel to the conveyor allowing for inspection and maintenance.

2) Installation of tachymetric rollers that stop the conveyor's motor if they do not turn.

3) Metallic particle detector (before milling, for instance).

4) Installation under a covered arcade or closed by panels in order to protect against rain and bad weather.

5) Dust is a problem for certain products (called dusty products) at the beginning and end of conveyor belts. Limestone, bauxite and phosphates are very dusty. Iron ore, coal and potash produce very little dust. In order to avoid the release of dust, falling product at the extremities of the conveyor must be regular, without jolts or spillage. Between the cover edge and the belt we must fit metal side panels extended by elastomer flaps. The conveyor cover can completely cover the entire length if it is depressurized (1–2 cm water column) by a ventilator discharging into a sleeve filter.

6) If the dust is explosive, the conveyor must run under an inert atmosphere.

2.3.5. *Choice of belt width*

When using a belt the first step is to choose the belt width.

Essentially, this width depends on the flowrate and can only be determined by trial and error. However, width must also satisfy the following conditions relative to the presence of larger fragments, i.e. fragments larger than 0.1 m:

– if the proportion of large fragments is significant:

$$\ell \geq 5d_M$$

d_M: the diameter of large fragments (m);

– with a mixture containing few large fragments:

$$\ell \geq 4d_M;$$

– if the large fragments are exceptional (sifted product):

$$\ell \geq 0.4 \left[\frac{d_M}{0.1} \right]^{2/3}.$$

2.3.6. *Flowrate*

Belt velocity is set according to the width selected and the type of product to be transported.

The velocity of the belt is given by the manufacturer. This velocity:

– increases with the width of the belt velocity;

– decreases with abrasivity (Mohs hardness).

Velocity decreases with product abrasiveness in order to minimize conveyor wear.

It increases with flowability of the product, as products with higher flowability present a decreased load per square meter of the belt. As we will see, the overload angle γ is almost half the angle of repose, which decreases with flowability.

Let us consider the general example of a trough conveyor with three identical rollers of length $\ell/3$, where ℓ is the belt width.

The loaded length of a lateral roller is:

$$\ell_c = 0.945\frac{\ell}{3} - 0.02 \quad (\text{in m})$$

Figure 2.4. *Cross-section of the product bed on a belt*

The cross-section (A_u) of the product bed (Figure 2.4) is the sum of the surfaces of a triangle and a trapezium:

$$A_u = \left[\frac{\ell}{6} + \ell_c\cos\beta\right]^2 tg\gamma + \left[\frac{\ell}{3} + \ell_c\cos\beta\right]\ell_c\sin\beta$$

γ: the overload angle (varying between 10° and 30°);

β: the inclination angle of rollers (between 20° and 30°).

If the conveyor's slope ascends from the horizontal axis, the angle of this slope α must not exceed the value of the overload angle γ if we wish to avoid significant product collapse.

The manufacturer knows the value of the overload angle γ corresponding to each product. This angle is inferior to the angle of repose.

Finally, the flow expression becomes:

$$Q = A_u V_B (1 - 0.0066\alpha)$$

Q: the flow volume ($m^3.s^{-1}$);

A_u: the cross-section of the product bed (m^2);

V_B: the conveyor velocity ($m.s^{-1}$);

α: the angle of the ascending slope to the horizontal axis: sexagesimal degrees.

2.3.7. *Power consumption*

The driving shaft power is:

$$P_a = gM_T V_B F_R C_S 10^{-3} \quad (kW)$$

where M_T is the total mass in movement (kg).

The linear density of the mechanical parts in movement is:

$$M_\ell = \dot{M}_M \left[\frac{\ell}{0.5} \right]^{0.8} + M_C \ell$$

ℓ: the belt width;

$12 \text{ kg.m}^{-2} < M_M < 25 \text{ kg.m}^{-2}$ (equivalent mass of rollers);

$10 \text{ kg.m}^{-2} < M_C < 20 \text{ kg.m}^{-2}$ (surface mass of belt).

The linear mass of the product is:

$$A_u \rho_a$$

The total mass in movement is:

$$M_T = (M_\ell + A_u \rho_a) L$$

F_R is the friction coefficient corresponding to the rollers' rotation:

$$0.025 < F_R < 0.03$$

Coefficient C_S is a safety coefficient set according to the length of the conveyor.

L < 50 m $C_S = 3$

50 m < L < 200 m $C_S = 3.35 - 0.011\,L$

L > 200 m $C_S = 1 + \dfrac{70}{L}$

NOTE.–

Motor power is P_a/η

$\eta = 0.95$ for a chain or cylindrical gears

$\eta = 0.70$ for a worm gear unit

EXAMPLE.–

$\ell = 1$ m $M_M = 17$ kg.m^{-2} $M_C = 15$ kg.m^{-2} $V_B = 2.3$ m.s^{-1}

$\beta = 20°$ $\gamma = 15°$ L = 200 m $\rho_a = 1{,}200$ kg.m^{-3}

$$\ell_c = \frac{0.945}{3} - 0.02 = 0.295 \text{ m}$$

tg $\gamma = 0.268$ sin $\beta = 0.342$ cos $\beta = 0.934$

$$\left[\frac{1}{6} + 0.295 \times 0.934\right]^2 0.268 + \left[\frac{1}{3} + 0.295 \times 0.934\right] 0.295 \times 0.342 = A_u$$

$$A_u = 0.052 + 0.061 = 0.113 \text{ m}^2$$

$$Q = 0.113 \times 2.3 = 0.26 \text{m}^3.\text{s}^{-1}$$

$$M_\ell = 17 \times 2^{0.8} + 15 = 44.6 \text{ kg.m}^{-1}$$

$$C_S = 1 + \frac{70}{200} = 1.35$$

Supposing:

$$F_R = 0.027$$

$$P_a = 9.81(44.6 + 0.113 \times 1,200) \times 200 \times 2.3 \times 0.027 \times 1.35 \times 10^{-3}$$

$$P_a = 29.64 \text{ kW}$$

2.3.8. *Effort incurred by the belt*

Let us write the friction balance equation from pulley theory:

$$T_1 = e^{\phi\alpha}T_2 \qquad\qquad [2.1]$$

ϕ is the friction coefficient of the belt on the roller;

$\phi = 0.25$ is the dry rubber belt on the steel roller;

$\phi = 0.35$ is the rubber belt on the rubber roller.

We will now write the balance of forces:

$$T_1 - T_2 = T_E \text{ (working tension connected to shaft power } P_a) \qquad [2.2]$$

$$T_E = \frac{1,000 P_a}{10V} = \frac{100 \ P_a}{V} \quad (\text{daN})$$

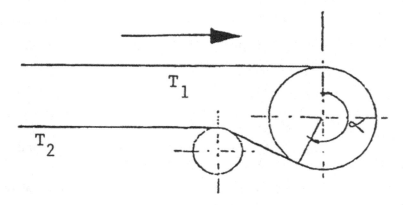

Figure 2.5. *Belt tension*

From relationships [2.1] and [2.2], we conclude:

$$T_1 = T_E \frac{e^{\phi\alpha}}{e^{\phi\alpha} - 1} \qquad T_2 = T_E \frac{1}{e^{\phi\alpha} - 1}$$

Tension T_2 is provided by the counterweight whose mass is $2T_2$ in kg.

Tension T_1 is the maximum tension that the belt is subjected to. The number of reinforcing folds must be given according to this value, which varies between 2 and 6. The acceptable stress arising from tension T_1 varies from 10 to 100 daN per centimeter of belt length.

EXAMPLE.–

$\phi = 0.25 \qquad \alpha = 180° = 3.14$ rad $\qquad V = 1$ m.s^{-1}

$Pa = 30$ kW $\quad \ell = 1$ m $= 100$ cm

$T_E = 3,000$ daN

$$T_1 = 3,000 \times \frac{2.19}{2.19 - 1} = 5,520 \text{ daN}$$

Hence, the stress:

$$\frac{T_1}{\ell} = 5 \times \frac{520}{100} = 55.2 \text{ daNcm}^{-1}$$

2.3.9. Loading systems

A correctly loading design should allow for:

– placing all of the required product on the belt without losses and with a sufficient density per m^2 to allow for the minimum belt velocity;

– limiting belt wear arising due to impacts by means of "soft" arrival of product on the belt.

To this end, we must:

– distribute the product at a regular flowrate, and more precisely at a velocity close to that of the belt's running velocity;

– avoid dust fly off.

One solution is to use an inclined chute whose slope and manufacture material are listed in Table 2.1.

Material transported		Maximum chute slope	Chute manufacture material
Cereals		25–30°	Ordinary steel
Coal			
	Mixed	40–45°	Steel
	Gauged	30–40°	Steel
	Fines	50–60°	Stainless steel (or plastic coated)
Coke			
	Mixed	25–30°	Extra-hard coating
	Fines	30–35°	Corhart, cast basalt, etc.
Ores			
	Mixed	45°	Manganese steel
Large caliber		40°	Rubber
Fine		50°	
Dried fertilizers		35°	Ordinary steel
Sticking		40°	Stainless steel
Phosphates		60°	Steel

Table 2.1. *Maximum slope and chute material*

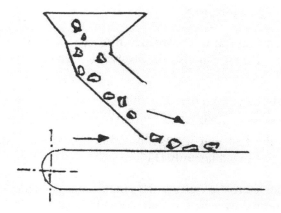

Figure 2.6. *Schema of a feeding chute*

Chute width is somewhat less than that of the belt, while vertical size is the result of a compromise:

– to avoid arching;

– to provide a flowrate limited to the desired value.

If the flow comes from a hopper, the hopper can be fitted with a slit-shaped aperture and the vertical distance from the mouth to the belt will increase moving downstream (Figure 2.7). Thus, flow is uniform at the lower part of the hopper.

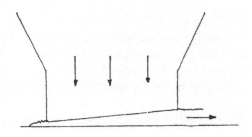

Figure 2.7. *Feeding via a slit under a hopper*

Other systems are described in connection with hoppers and silos.

2.3.10. *Discharge systems*

Discharge can occur at the downstream extremity by simply falling after the entrainment roller. It can also be ensured at a given point selected along the conveyor by:

– a lateral discharge scraper. For this, the belt must be set flat and the scraper positioned diagonally;

– a bend in the belt and lifting discharge up. Tipped into a chute, the product is then evacuated laterally (Figure 2.8);

– a raising device to the right *or* the left, allowing for a pile to be made to either side of the conveyor belt and all along its length.

Figure 2.8. *Tripper car*

2.3.11. *Metal conveyor belt*

Typically, these devices are used for hot products or for products that must undergo thermal treatment.

Belt thickness is between 0.4 and 1.4 mm, which is somewhat higher than that of the sheets of plate heat exchangers. The typical widths, which are selected according to the flow required, vary according to the case between 0.2 and 1.5 m. The belt can be supported by rollers, wheels, wooden or plastic slides, or by a metal table if a thermal transfer is applied.

Metal belts, which are always flat (never trough shaped), transport flow volume:

$$Q = \eta_v A V_B$$

V_B: the running velocity of the belt $(m.s^{-1})$;

Q: the volume flowrate $(m^3.s^{-1})$;

η_v: the filling ratio;

A: the hypothetical transversal cross-section of the product bed (supposing the cross-section is square) (m^2), with

$$A = \ell^2$$

ℓ: the belt width (m).

When the product is flowing, the volumetric yield η_v decreases significantly, thereby increasing the risk of it being lost due to overflow. Here, we write:

$$\eta_v = \frac{1}{n}$$

Flowability class		n
A	high flowability	13–16
B	low flowability	8–11
C	no flowability	3–7

We should also add pastes of low consistency into class A (fluid honey, thick syrup), while class C contains wet fines and ultrafines ($d_p < 20$ μm).

An ascending inclination of $10°$ towards the horizontal axis decreases the capacity by approximately 12% and an inclination of $20°$ decreases it by 25%.

Products subjected to thermal treatment can receive heat from radiation above the belt or conduction through the belt. In both cases, we should adapt the surface load (kilograms of product per square meter of belt) ensuring the thermal transfer provides the required thermal transfer surface according to the mass to be treated. The surface load is given by:

$$M_s = \frac{\eta_v A \rho_a}{\ell_u}$$

ρ_a: the apparent density of the product (kg.m^{-3});

ℓ_u: the real loaded width of the belt (m), with

$$0.75 \, \ell < \ell_u < 0.8\ell;$$

ℓ: the total belt width (m).

2.4. Chain conveyor(s)

2.4.1. Crossbar conveyor

The crossbar conveyor was studied by Korzen [KOR 91a, KOR 91b], for both horizontal and vertical conveyors.

An endless chain carries the crossbars, sliding to the end of a trough of rectangular cross-section. The lower side (active part of chain) enters the trough, with the crossbars entraining the product. On exiting the trough, the product leaves the conveyor by falling between the crossbars.

The chain then goes onto the end wheel, becoming the return side (top). These chains are robust, being capable of accepting an effort of several force-tons. Simple and inexpensive sliding chains consume more energy than those carried by rollers.

This conveyor can be covered, which helps to avoid the release of dust. The flow of crossbar conveyors is:

$$Q = \eta_v \ell^2 V_c$$

Q: the volume flowrate of the product (m^3.s^{-1});

ℓ: the trough width (m), with

$$0.1 \text{ m} < \ell < 1 \text{ m};$$

V_c: the chain velocity (m.s^{-1}).

Velocity V_c decreases with product abrasiveness, i.e. with its hardness (see Appendix 1).

Abrasiveness class	V_c
I	$0.5 \ \text{m.s}^{-1} < V_c < 1 \ \text{m.s}^{-1}$
II	$0.2 \ \text{m.s}^{-1} < V_c < 0.5 \ \text{m.s}^{-1}$
III	$V_c < 0.2 \ \text{m.s}^{-1}$

For η_v the volumetric yield for a hypothetical trough of square cross-section with side ℓ, the yield decreases with flowability.

Flowability class	η_v
A – high flowability	0.10
B – low flowability	0.25
C – no flowability	0.40

Power when empty (in kW) is:

$$P_1 = 10^{-3} M_c L n_c V_c \ \cos\alpha \ g$$

L: the conveyor length (m);

n_c: the number of parallel chains;

α: the inclination angle (ascending or descending) to the horizontal axis;

V_c: the chain velocity (m.s^{-1});

M_c: the linear density of the chain (kg.m^{-1}).

For reference purposes:

Trough width ℓ (m)	M_c (kg.m^{-1})	Breaking strength (daN)
0.25	3	3,000
0.35	10	13,000
	17	27,000

Power connected to the product is:

$$P_2 = 10^{-3} gV_c \ell^2 L\rho_a \left(\cos\alpha\, tg\phi_p + \sin\alpha\right)$$

ϕ_p : the friction angle of the product at the bottom of the trough.

For example, with a steel trough:

Product	$tg\,\phi_p$
Lean coal	0.33
Flour, fats, wood shavings, coke	0.36
Fat coal, limestone	0.58

EXAMPLE.–

$\ell = 0.5$ m	$V_c = 0.3$ m.s^{-1}	$\eta_v = 0.25$
$M_c = 15$ kg.m^{-1}	$L = 40$ m	$n_c = 1$
$\alpha = 20°$	$\rho_a = 1,200$ kg.m^{-3}	$tg\,\phi_p = 0.36$

$$Q = 0.25 \times (0.5)^2 \times 0.3 = 0.01875 \text{ m}^3.\text{s}^{-1}$$

$$Q = 67.5 \text{ m}^3.\text{h}^{-1}$$

$$P_1 = 10^{-3} \times 15 \times 40 \times 1 \times 0.3 \times 0.94 \times 9.81 = 1.66 \text{ kW}$$

$$P_2 = 10^{-3} \times 9.81 \times 0.3 \times (0.5)^2 \times 40 \times 1,200 (0.94 \times 0.36 + 0.342) = 24 \text{ kW}$$

Shaft power:

$$P_a = 24 + 1.66 = 25.66 \text{ kW}$$

2.4.2. Scraper conveyors

Unlike in crossbar conveyors, (vertical) scrapers transport the product perfectly even when it is flowing, and moreover the device works correctly at inclinations of up to 40°.

The conveyor is designed with two parallel chains to operate the scrapers without leading to any warping.

The expressions concerning flow and power are identical to those of the crossbar conveyor, with one difference: here yield $\eta_v = 1$.

The table below provides the values that characterize these chains:

M_c (kg.m^{-1})	Breaking strength (daN)
12	7,000
23	15,000

The tractive effort admissible is taken as being 15% of the breaking strength.

2.4.3. Metal pallet conveyors

This type of conveyor is used for hot products (ingots in metallurgy).

The empty power is given by:

$$P_1 = 10^{-3} \, g V L M_\ell \left(tg\phi_c + tg\phi_V \right),$$

ϕ_c and ϕ_V : the friction angles of chains for loaded sides and returning empty sides, respectively, and:

$$tg\phi_c = 0.3 \quad tg\phi_V = 0.16$$

Two or three parallel chains can be used.

Linear density (M_ℓ) is:

$$M_\ell = n_c M_c + M_p$$

M_c: the linear density of chains (kg.m^{-1});

n_c: the number of parallel chains;

M_p: the mean linear density of pallets (kg.m^{-1}), with:

$$20 \text{ kg.m}^{-1} < M_c < 45 \text{ kg.m}^{-1}$$

$$40 \text{ kg.m}^{-1} < M_p < 90 \text{ kg.m}^{-1}$$

Power connected to the product (in kW) is given by:

$$P_2 = 10^{-3} \text{gLM}_p \text{tg} \phi_c V_c$$

M_p: the mass of the product per linear meter of the conveyor (kg.m^{-1}).

Typically, metal pallet conveyors are used horizontally.

NOTE.–

The influence of mechanical properties of the divided solids was studied by Korzen [KOR 91a, KOR 91b]. These characteristics are significant for scaling the device.

2.5. Vibrating conveyor

2.5.1. Operational principle

This conveyor is made up of a trough supported by flexible leaf springs (Figure 2.9). The trough can be made to vibrate by several different systems:

– electromagnet;

– eccentric shaft;

– crank.

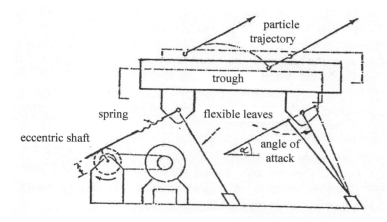

Figure 2.9. *Twin mass and guided oscillation conveyor*

The particles perform successive jumps. When they leave the trough, they are at an angle α to the horizontal axis, called the angle of attack. This angle is the same as that made by the flexible leaves to the vertical.

Energy is required to maintain the vibrations, as there is inevitable amortization. This amortization stems from:

– the mechanical parts of the device. This damping applies even if the trough is empty;

– the product conveyed. The fall into the trough occurs by inelastic shocks. In addition, there is friction between the product and the bottom and sides of the trough, together with internal friction in the product itself.

The mechanical stress is proportional to the amplitude of these vibrations and the square of pulsation. This is why vibrating feeders, which are smaller, can tolerate high frequencies, which is not the case for transporters.

2.5.2. *Types of conveyors*

2.5.2.1. *Direct action (single mass)*

The vibration generator is rigidly attached to the conveyance pathway. The whole setup can vibrate as it is attached to the surrounding structure by springs or rubber shock absorbers. The oscillations are then referred to as free. If the transporter is placed on elastic blades that are slightly vertically inclined, the oscillations are referred to as guided.

Vibrations are produced by an electromechanical device, i.e. a motor that leads a crank and rod system, an unbalanced shaft, or an eccentric shaft.

The operational frequency is higher than the natural (resonance) frequency of the system.

Flow is regulated by action on the angle of attack and the vibration frequency (velocity of the unbalanced shaft or the crank and rod system). Only low precision is obtained, since the mass of product present in the trough has a bearing on the vibration amplitude.

These devices are noisy. They are suited to both high flowrates and significant lengths.

2.5.2.2. Indirect action (dual mass)

The vibration generator is elastically connected to the pathway (springs, metal or reinforced plastic blades) and the pathway is elastically attached to the structure too.

The equivalent mass of the whole setup is a function of the attachment spring's stiffness and the respective masses of the generator and the pathway.

Flow is regulated, as above, by action on the angle of attack. We can also modify the mass of the generator; however, this should be considered as part of the initial calibrations that cannot subsequently be changed.

Flow regulation is performed as follows:

– electromagnetically: N = 50 Hz, A = 1 mm, n acts on the feed tension (flowrate regulation is easy from 0% to 100%).

– electromechanically: 12, 16 or 25 Hz, A = 0.5 mm, we modulate motor velocity (flow regulation occurs from 5% to 100%).

Electromagnetic electricity consumption is significantly lower than the electromechanical counterpart.

2.5.3. *Progression rate of the product in the trough*

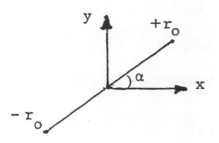

Figure 2.10. *Displacement of all the points in the trough*

$$\text{Stroke} = 2A\,(A = \text{amplitude})$$

One particle follows the surface via the climbing phase through which it acquires velocity and presents a parabolic trajectory. Through the descending phase, the particle comes into contact with the trough until it is projected the next time.

The oblique displacement is expressed by:

$$r = r_o \sin\omega\left(t + t_o\right) = r_o \sin\left(\phi + \phi_o\right)$$

with

$$\phi = \omega t \qquad\qquad \phi_o = \omega t_o$$

Velocity and acceleration are given by:

$$r' = \omega r_o \cos\omega\left(t + t_o\right) = \omega r_o \cos\left(\phi + \phi_o\right) \text{ and}$$

$$r'' = -\omega^2 r_o \sin\omega\left(t + t_o\right) = -\omega^2 r_o \sin\left(\phi + \phi_o\right), \text{ respectively.}$$

If α is the angle of attack, the vertical acceleration of the trough is:

$$y'' = -\omega^2 r_o \sin\alpha \, \sin\left(\phi + \phi_o\right)$$

As vertical acceleration of the trough is directed downward (y" < 0) and its absolute value exceeds the acceleration of gravity g, the product rises relative to the trough. At this instant:

$$g = \omega^2 r_o \, \sin\alpha \, \sin(\phi + \phi_o) \qquad [2.3]$$

Now, let us take this instant as the start time:

$$\omega t = \phi = 0$$

In addition, writing:

$$k = \frac{g}{\omega^2 r_o \, \sin\alpha} \qquad [2.4]$$

Equation [2.3] is written as:

$$\sin\phi_o = k < 1 \qquad [2.5]$$

The particle rising occurs at time $t = \dfrac{\phi_o}{\omega}$.

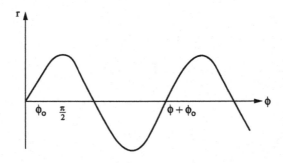

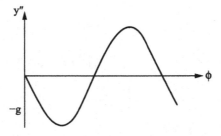

Figure 2.11. *Development of oblique displacement r and vertical acceleration y"*

At this precise moment, the particle is still resting on the trough, and its coordinates are:

$$y_a^{(o)} = r_o \, \sin\alpha \, \sin\phi_o$$

$$x_a^{(o)} = r_o \, \cos\alpha \sin\phi_o$$

Velocity at this moment is:

$$y'_p = y'_a = \omega r_o \, \sin\alpha \, \cos\phi_o$$

$$x'_p = x'_a = \omega r_o \, \cos\alpha \, \cos\phi_o$$

Starting from this initial position, the particle presents a parabolic trajectory, whose parametric equations are:

$$y_p = r_o \, \sin\alpha \, \sin\phi_o + \omega t r_o \, \sin\alpha \, \cos\phi_o - \frac{1}{2}gt^2$$

$$x_p = r_o \, \cos\alpha \, \sin\phi_o + \omega t r_o \, \cos\alpha \, \cos\phi_o$$

Simultaneously, the vertical axis for the whole of the trough follows the law:

$$y_a = r_o \, \sin\alpha \, \sin(\phi + \phi_o)$$

and the horizontal axis from the lifting point is:

$$x_d = r_o \, \cos\alpha \, \sin(\phi + \phi_o)$$

The product descends to the trough when:

$$y_p = y_a$$

that is;

$$r_o \, \sin\alpha \, \sin\phi_o + \phi r_o \, \sin\alpha \, \cos\phi_o - \frac{g\phi^2}{2\omega^2} = r_o \, \sin\alpha \, \sin(\phi + \phi_o)$$

Now, dividing both members by $r_0 \sin\alpha$:

$$\sin\phi_o + \phi \, \cos\phi_o - \frac{\phi^2 \, \sin\phi_o}{2} = \sin\left(\phi + \phi_o\right)$$

that is:

$$\cot g\phi_o = \frac{\dfrac{\phi^2}{2} + \cos\phi - 1}{\phi - \sin\phi} \qquad [2.6]$$

During the vibration period, for the trough to be able to re-accelerate the descended particles, it is necessary that:

$$3\pi / 4 < \phi_o + \phi < 2\pi \quad \left(\text{e.g. } \phi = 1.75\pi\right)$$

From this, we can deduce the value of ϕ_o by equation [2.5] together with the value of amplitude r_0 by equation [2.3]. We have set frequency N of vibration (thus, we can deduce pulsation $\omega = 2\pi N$).

Consequently, the duration of the particles' free trajectory is:

$$\tau = \phi / \omega$$

The rate of progression of the solid through the trough can then be deduced as:

$$V_p = \frac{x_p - x_a}{\tau} = \frac{r_o \, \cos\alpha}{\tau}\left[\phi \, \cos\phi_o + \sin\phi_o - \sin\left(\phi + \phi_o\right)\right]$$

The rate of progression of the product can also be calculated by the following formula:

$$V_p = 0.03 \, N \, C$$

N: the frequency (mn^{-1});

C: the stroke (double the vibration amplitude) (m).

Thus, if we have a high frequency, we can accept a low stroke and *vice versa*. Both of these parameters vary inversely. In practice, we place them in the following intervals:

Feeders	5,000 To 1, 000	N	cy.mn^{-1}
	1.3 To 8	C	mm
Transporters	1,000 To 150	N	cy.mn^{-1}
	8 To 25	C	mm

In general, the excitation frequency N is different to the device's own frequency N_o:

– if $N > N_o$ operation is supercritical;

– if $N < N_o$ operation is subcritical.

2.5.4. *Flowrate*

$$Q = \ell \, hV_p$$

V_p: the rate of progression (m.s^{-1});

ℓ: the trough width (m);

h: the height of the product bed (m).

The bed height h depends on the angle of attack and the product (granulometry, humidity).

The real value of the rate of progression is lower than the theoretical value as impacts are not elastic and they produce friction between both the product and trough and between the particles themselves.

With powders, which are typically well aerated, the interstitial air acts as a pneumatic amortizer and the rate of progression decreases quickly with an increase in thickness h. In practice, a thin layer of product progresses over a compacted and practically immobile bed of powder. Thus, we should limit h to 1–3 cm and frequency N to low values.

A limited angle of attack α benefits horizontal acceleration, which reduces compacting and the possibility of clogging. Indeed, there is an optimal value for α.

2.5.5. *Power consumption*

A simple formula for a system close to resonance is:

$$P = 3.910^{-5} \, M \, N^3 C^2 \, \cos\phi_o$$

P: W;

M: the sum of mobile mass (kg);

N: the frequency (mn^{-1});

C: the stroke (m).

$$C = 2r_o$$

The mass of the product present in the trough is:

$$M_P = L_G \times h \times l \times \rho_a$$

In order to obtain the sum of mobile masses M, we should add the mass of the vibrating parts.

EXAMPLE.–

$\rho_a = 1,200 \text{ kg.m}^{-3}$ $L_G = 2 \text{ m}$ $N = 50 \text{ Hz}$

$\ell = 0.75 \text{ m}$ $h = 0.03 \text{ m}$ $\alpha = 30°$

$M = 150 \text{ kg}$

Supposing that:

$\phi = 1.75\pi = 5.5 \text{ rad},$

$$\cot g\phi_o = \frac{15.125 + 0.70867 - 1}{5.5 + 0.7055}$$

$$= \frac{14.83367}{6.2055} = 2.39040$$

$$tg\phi_o = 0.418339$$

$$\phi_o = 22°,7 = 0.3962 \text{ rad} \qquad\qquad \phi + \phi_o = 5.8962 \text{ rad} < 2\pi$$

$$\sin\phi_o = 0.38592 \qquad\qquad \cos\phi_o = 0.922528$$

We have:

$$\omega^2 r_o = \frac{g}{\sin\alpha \ \sin\phi_o} = \frac{9.81}{0.5 \times 0.38592} = 50.8395$$

$$\omega = 2\pi \times 50 = 314 \text{ rad.s}^{-1}$$

$$r_o = \frac{50.8395}{(314)^2} = 0.5.10^{-3} \text{ m}$$

The instant of maximum vertical trajectory is:

$$t_{max} = \frac{\omega r_o \ \sin\alpha \ \cos\phi_o}{g} = \frac{314 \times 0.5.10^{-3} \times 0.5 \times 0.922528}{9.81}$$

$$t_{max} = 0.007382 \text{ s}$$

The instant of "landing" is:

$$\tau = \frac{5.5}{314} = 0.01752 \text{ s} > t_{max}$$

$$\frac{y_{pmax}}{r_o} = 0.5 \times 0.38592 + 314 \times 0.007382 \times 0.5 \times 0.922528 - \frac{9.81}{2} \times (0.007382)^2$$

$$= 0.19296 + 1.06918 - 0.000267$$

$$\frac{y_{pmax}}{r_o} = 1.262407$$

The solid's progression rate is given by:

$$V_p = \frac{10^{-3}}{0.01752} \times 0.866 \left(5.5 \times 0.922528 + 0.38592 + 0.37945 \right)$$

$$V_p = 0.144 \text{ m.s}^{-1}$$

So, $C = 2r_0 = 10^{-3}$ m; consequently:

$$V_p = 0.03 \times 60 \times 50 \times 10^{-3}$$

$$V_p = 0.09 \text{ m.s}^{-1}$$

Hence, kinetic yield is given by:

$$\eta_c = \frac{V_{p\ empirical}}{V_{p\ theoretical}} = \frac{0.09}{0.144} = 0.62$$

$$Q = 0.03 \times 0.75 \times 0.09 = 2.10^{-3} \text{ m}^3.\text{s}^{-1}$$

The mass of the product present in the trough is:

$$M_p = 2 \times 0.03 \times 0.75 \times 1,200 = 54 \text{ kg}$$

We estimate the total vibrating mass at 150 kg as:

$$P = 3.9.10^{-5} \times 150 \times (3,000)^3 \times (10^{-3})^2 \times 0.9225$$

$$P = 146.4 \text{ W}$$

2.5.6. *Advantages and drawbacks of vibrating conveyors*

Vibrating conveyors are self-cleaning. They can manage hot products without any negative effect. Furthermore, they can be heated to limit the effects of humidity (adherence).

Electromechanical excitation is preferable for flows greater than 5 t.h^{-1}. This system is easy to maintain and can be made flameproof.

It is difficult to manage dusty products with these devices for two reasons:

– it is difficult to create effective seals against dust;

– a thin layer of fluidized product flows on an immobile bed of product compacted by vibration.

Lining the trough with an elastomer can help minimize the effects of adherence.

2.6. Elevators

2.6.1. *Bucket elevator principle*

A bucket elevator includes a metal chain or rubber belt onto which the buckets transporting the product are attached. This device is also known as a noria, a word of Spanish origin. A noria is used to take water from a well using an endless belt (rope) onto which recipients are attached.

Bucket elevators are used when the difference in level does not allow use of conveyor belt or chain conveyor (both crossbar and scraper). However, these elevators are to be avoided if the product is sticky.

2.6.2. *Types of elevators and discharge*

Bucket elevators are noted for their means of discharge, which is the means by which product is emptied from the top part of the device.

1) Centrifugal discharge:

In these devices, the spacing of buckets is equal to three times their opening. Bucket filling occurs by scooping.

If buckets are attached to a chain, the chain velocity is approximately 1.5 m.s^{-1}. If they are attached to a belt, the velocity can reach 2–4 m.s^{-1} (with cereals for instance).

The choice of velocity is determined by the radius of the head roller. Experience shows that discharge is acceptable when gravity balances the centrifugal acceleration:

$$g = \frac{V^2}{R}$$

Hence;

$$V = \sqrt{Rg}$$

where R is the radius of the head roller.

In practice, we can reduce the velocity to 70% of that indicated by the formula. In particular, we reduce velocity for products that are more abrasive than cereals, such as coal, coke, gravel and ore. If velocity is too high, the buckets wear out quickly during loading by scooping.

As a general rule, centrifugal discharge elevators are used for products that are easy to shovel, i.e. whose sizes are no greater than 5 cm. In addition to the products already mentioned, we may add sand, cement and clinker to the products treated this way.

2) Discharge by dumping:

Discharge occurs by turning the buckets over and simply letting the product fall into a chute. In devices such as these, the spacing of buckets is equal to 1.5 times their opening.

The running velocity is between 0.4 and 0.8 m.s^{-1}. This velocity is lower in order to neutralize the centrifugal force. The maximal value of 0.8 m.s^{-1} applies for chains.

The buckets are designed so that the back of each acts to guide the product contained in the next, but can also be emptied vertically from the top roller, and for this end, a pulley under the roller pushes the descending side against the rising side to clear the vertical space.

Discharge by dumping is also used for continuous elevators (whose buckets are joining). With a constant flow, the value of velocity is minimal, limiting the release of dust or occasioning larger pieces to bounce.

2.6.3. Methods of filling buckets

The buckets can be filled directly by taking product from a pile, that is, by being dragged (scooping) if the discharge is a centrifuge. With sufficient velocity, the product can be held in the bucket by inertia before the bucket reaches its position on the ascending side. This filling method is not recommended if the (fine or ultrafine) product flows easily.

Filling can also occur by the force of gravity under a chute, provided there is no empty space under the buckets when viewed horizontally.

2.6.4. Should a belt or chain be used?

Belts have the following advantages:

– higher velocity and therefore higher flowrate;

– quiet, regular and smooth operation;

– abrasion resistance (hardness above 7 on the Mohs scale);

– corrosion resistance (soda, wet salt, wet coal high in sulfur, etc.).

If the temperature exceeds 120 °C, a chain is required. This must be made of an abrasion-resistant material. In practice, its Brinell hardness B is connected to the M hardness of the product on the Mohs scale by the stress:

$$B \geq 150\, M - 350$$

Let us recall that the Brinell hardness of soft iron is 100 and that of hardened steel is 400. Appendix 1 gives the Mohs hardness of various products.

2.6.5. *Flowrate*

Flowrate (Q) is given by:

$$Q = R\Omega_G n_G V_G$$

Q: the flow ($m^3.s^{-1}$);

V_G: the running velocity of a bucket ($m.s^{-1}$);

Ω_G: the total volume of a bucket (m^3);

R: the filling ratio of buckets, with:

$0.5 < R < 0.85$, and typically $R \# 0.75$;

n_G: the number of buckets per meter of chain or belt (m^{-1}), with:

$$n_G = \frac{1}{p} = \frac{1}{e + L_G}$$

p: no buckets (m);

e: the space between buckets (m);

L_G: the size of buckets along the chain or the belt (m^{-1}).

Hence,

$$Q = R\frac{\Omega}{p}V_G$$

Recall that for:

– continuous elevators:

$$p = L_G$$

– dumping elevators:

$$p \# 1.5 L_G$$

– centrifugal discharge elevators:

$$2.5L_G < p < 3L_G$$

2.6.6. *Power consumption*

The mass to be lifted is:

M_G: the mass of a full bucket (kg);

n_G: the number of buckets per unit of length (height) (m^{-1});

H: the height difference (m).

Power corresponding to the difference in height is:

$$gM_G n_G HV_G$$

(This power covers the empty power P_1, the power due to the presence of the product P_2 and the power due solely to lifting the product itself P_3).

The force exerted for dragging is:

$$F_D = k[M_G D_T n_G g]$$

D_T: the diameter of the lower roller (m).

The quantity within the square brackets represents the weight of the product taken from the pile to the lower part of the elevator and k is a sort of friction coefficient:

– k = 0.67 for flowing products (in reality, low-rigidity products): aerated powders, cereals.

– k = 1 for products with low or no flowability (with high rigidity): fines + lumps.

If feeding occurs by means of an inclined channel and if, as is often the case in such circumstances, jolts are expected, we can use the same expression with a coefficient of k = 0.5, despite the absence of any dragging effort in this instance.

Finally, the shaft power is given by:

$$P_a = gV_G M_G n_G (H + kD_T)$$

2.6.7. *Vibrating elevator*

This looks like a spiral lane with a vertical axis. It does not consume much energy as it includes few moving parts, and it does not produce noise.

The product's retention time is high to limit floor congestion, which favors thermal exchanges:

– heating or drying by insufflation of air beneath the perforated lane;

– cooling by sprinkling water beneath the lane.

2.6.8. *Screw lifts*

Without an intermediate level, height is limited to around 15 m.

Flow can reach 200 $m^3.h^{-1}$ as the rotation frequency is high.

On feeding, a horizontal feeder screw is sometimes required.

Screw diameter (m)	Rotation frequency (rev.s^{-1})
0.1	7.5
0.2	4.2

2.7. Screw conveyor for divided solids

2.7.1. *General*

The screw conveyor is composed of a screw blade mounted on a shaft supported by bearings and turning in a trough.

The product fills the lower part of the trough and rubs against its lining. It is then pushed by the movement of the screw slipping through the product. This device is inexpensive and is easy to maintain.

It can also be used to mix powders. In such cases, the screw is replaced by paddles.

We can also use a screw if the pathway is less than 100 m and if there is not much room and if we desire good dust seal. Depressurizing it with a 5–10 cm water column can help eliminate odors if the product is foul-smelling.

The product must not be sticky.

2.7.2. Choice of screw characteristics

2.7.2.1. Screw diameter

This depends on the fragment size and the mass content of the product. The table below gives the maximum sizes of the lumps in millimeters according to screw diameter D.

D (mm)	100	150	200	300	400	500	600
20–25% of lumps	10	20	40	50	80	90	90
100% of lumps	5	10	20	25	40	50	60

According to lump size, clearance between the screw and the launder can vary between 5 and 20 mm.

When the lump content is high, clearance between the screw blade and the launder must be at least equal to the size of these lumps in order to avoid blocking the screw.

2.7.2.2. Screw thread

The thread pitch lies between 0.5D and 1.5D and is often equal to D.

A large thread is compatible with a product of high flowability.

The greater the screw incline, the greater the reduction in pitch and increase in velocity. A pitch of size 0.5D is recommended for screw feeders.

2.7.2.3. Rotation frequency

Due to abrasion, rotation frequency decreases as screw diameter increases. We will write this to the nearest 5%:

$$N = 76 - 60D$$

N: the rotation velocity (rev.min^{-1});

D: the screw exterior diameter (m).

2.7.3. Darnell and Mol's theoretical calculations for screw flow [DAR 56]

The solid behaves like an elastic cord with rectangular cross-section. These authors propose a calculation to establish the maximum flowrate of the screw.

Here, the plane of reference is any given plane that is perpendicular to the screw axis, where:

– ϕ: the screw angle with plane of reference perpendicular to the screw axis;

– h: the depth of the screw thread (perpendicular to axis) (m).

Screw angle ϕ increases as we approach the axis. The authors selected the corresponding value at distance h/2 from the bottom of the thread.

The authors used the force torque relative to the screw axis.

D is the barrel diameter (and also screw diameter) (m), with:

$$E = \frac{D-h}{D} \qquad K = \frac{E(\mathrm{tg}\phi + \mu_v)}{1 - \mu_v \mathrm{tg}\phi} \qquad C = \frac{D-2h}{D}$$

K is defined as the ratio:

$$\frac{\text{force perpendicular to the plane of reference}}{\text{force parallel to the plane of reference}}$$

μ_v: the friction coefficient of the cord on the screw metal;

$$M = C(K\sin\phi + C\cos\phi) + \frac{2h}{p}(KCtg\phi + E^2) + \frac{hE}{L\mu_f}\sin\phi(E\cos\phi + K\sin\phi)Ln\frac{P_2}{P_1}$$

L: the axial length of the screw (m);

μ_f: the friction coefficient of the bed on the barrel;

p: the screw pitch (m);

P_1 and P_2: the upstream and downstream pressures of the divided solid cord (Pa).

The motion angle θ is the angle made by the displacement of the cord relative to the plane of reference. The authors found that:

$$\cos\theta = K\sin\theta + M(\phi, P_2 / P_1, h)$$

Finally, the theoretical volume flowrate Q of solid delivered by the screw is:

$$\frac{Q_T}{N} = \frac{\pi^2 Dh(D - h)tg\theta tg\phi}{tg\theta + tg\phi} \qquad \text{([DAR 56] equation 13)}$$

N: the rotation velocity of the screw (rev.s^{-1}).

The authors studied the influence of screw geometry on angle θ and flow Q_T. They provided a numerical example for calculating Q_T.

We must also apply the trough's filling ratio. Let this ratio be η_R.

If the product is cohesive, it will tend to turn in unison with the screw and not advance. The product must have maximum flowability order to achieve satisfactory flow.

Volumetric yield η_v can be estimated in the following manner, as shown in Table 2.2.

Flowability class		η_v
Perfect	A	0.8
Average	B	0.5
Low	C	0.4

Table 2.2. *Volumetric yield*

To prevent the product from rotating in unison with the screw, it is necessary for the screw threads to be smooth and trough lining to be rough.

In order to increase flow, we can increase pitch p; however, this is only possible if the product has sufficient flowability. If the product is cohesive, one solution can be to position 2 parallel screws that would turn in the same direction so that the threads can penetrate one another. This system prevents the product from rotating in unison with the screw.

If a product proves to be too cohesive, we can prevent it from turning in unison, though without the possibility of increasing flow, by replacing the full screw thread with a ribbon that shears the product.

An upward inclination of angle α to the horizontal axis is announced by a reduction in flowrate that can be shown, on average, by multiplying it with the coefficient:

$$(1 - 0.02\alpha) \hspace{3cm} (\alpha: \text{sexagesimal degrees})$$

Rautenbach and Schumacher [RAU 87] provided an empirical expression of flow. They applied Jenike's function, which we will denote k_E (see section 1.3.7 [DUR 16a]).

Finally, the flowrate becomes:

$$Q = Q_T \eta_R \eta_v (1 - 0.02\alpha)$$

with:

$$Q_T = 0.5N \, p \frac{\pi D^2}{4}$$

Mass flowrate is:

$$W = \rho_a Q,$$

ρ_a: the apparent density (kg.m^{-3}) (see Appendix 2).

2.7.4. Screw power consumption

This power is the sum of the empty power and the power required when product is present.

The empty power is calculated according to American firm CEMA's formula as:

$$P_1 = 5.8.10^{-3} \, LNF_D F_P D \ (kW)$$

L: the screw length (m);

N: the rotation frequency (rev.s^{-1}).

Coefficient F_D is subject to the screw diameter:

D (mm)	F_D
≤ 300	3
$300 < \phi \leq 400$	5
$400 < \phi \leq 600$	8

Coefficient F_P depends on the type of bearing:

Type of bearing	F_P
Ball bearings	2
Plain bearings (bushes)	4

Power due to friction in the product is again calculated according to the CEMA as:

$$P_2 = 19.3.10^{-3} \ LQ\rho_a F_S n_F F_V \ (kW)$$

Q: the volume flowrate ($m^3.s^{-1}$);

ρ_a: the apparent density of the product ($kg.m^{-3}$);

n_F: the number of screw threads;

F_V: the coefficient equal to 1 if the screw is classic and equal to 2 if it is equipped with paddles and non-continuous threads;

F_S: the solid characteristic coefficient (see below).

Product elevation consumes power P_3:

$$P_3 = gW \ H.10^{-3} \ (kW)$$

H: the height difference (m)

W: the mass flowrate of the product ($kg.s^{-1}$);

g: the acceleration due to gravity ($9.81 \ m.s^{-2}$).

Coefficient F_S characterizes the resistance to product progression. The table below proposes an estimate for this coefficient.

Flowability class	F_S
A	1
B	2
C	3

In a similar manner to their work on volume flowrate, Rautenbach and Schumacher [RAU 87] established an empirical expression for torque resisting rotation by applying product flowability, that is, ratio k_E, which they called ff_r.

NOTE.–

Screw wear is subject to product abrasiveness, and screw life expectancy can be divided by 3 or a higher number for harder products. Appendix 1 lists the Mohs hardness of various products together with the definition of hard and soft products.

EXAMPLE.–

$N = 1.17$ rev.s^{-1} $D = 0.6$ m $F_D = 8$

$\rho_a = 1,200$ kg.m^{-3} $L = 10$ m $F_P = 4$

$\eta_v = 0.3$ $\alpha = 20°$ $F_S = n_F = F_V = 1$

$$Q = 0.3 \times (1 - 0.02 \times 20) \times 1.17 \times 0.6 \times \frac{\pi}{4} \times 0.6^2 = 0.04 \text{m}^3 . \text{s}^{-1}$$

$$P_1 = 5.8.10^{-3} \times 10 \times 1.17 \times 8 \times 4 \times 0.6 = 0.16 \text{ kW}$$

$$P_2 = 19.3.10^{-3} \times 10 \times 0.04 \times 1,200 = 9.25 \text{ kW}$$

$$H = 10 \sin 20° = 3.42 \text{ m}$$

$$P_3 = 9.81 \times 0.04 \times 1,200 \times 3.42 \times 10^{-3} = 1.61 \text{ kW}$$

Hence, shaft power is given by:

$$P_a = 0.16 + 9.25 + 1.61 = 11.02 \text{ kW}$$

NOTE.–

Relevant, practical data concerning conveyor screws can be found in the following publications:

– Bates [BAT 69];

– Burkhardt [BUR 67];

– Rehkugler [REH 67].

2.8. Transport by loads

2.8.1. *Containers and skips*

Here, the product is placed in capacities that are:

– either containers smaller than 5 m^3;

– or skips (road or rail) between 5 and 20 m^3 in volume.

These capacities are emptied by either gravity or a gas flow. This can occur:

– from above;

– from below or the sides.

2.8.1.1. *From above*

Tilting and tipping along the length of one side. Segregation is not significant and there is no danger of blockage.

Pneumatic suction of flowing products, avoids the release of dust when the flowrate is limited and segregation is difficult to avoid.

2.8.1.2. *From below or the sides*

Opening a trap below or along one side of a skip. This leads to a general flow with little segregation.

Having an aperture of diameter < 0.25 m below a container. Emptying occurs with the creation of a pit according to product cohesion. This discharge is segregative.

Pneumatic suction of flowing products from below (limited flow, division possible, no dust).

Container (and, moreover, skip) filling can be automated and controlled by means of a sensor that measures the top level.

2.8.2. *Roller conveyor for boxes*

Boxes are conveyed on a series of free rollers.

For objects of average density, the displacement velocity must be >0.4 m.s^{-1} if regular movement is desired. In practice, a 3% slope leads to a velocity close to 1 m.s^{-1}.

Here, we will not address palletization techniques for handling and storing crates, packets, rolls, etc.

2.8.3. *Material lifters*

The most basic of these devices would a basket lifted by a rope via a pulley. This process is still used in certain sites and in certain Greek monasteries.

The crate, guided and balanced by a counterweight, is the elevator itself. The counterweight mass is equal to that of the empty box with the addition of a half charge.

Two crate elevators, balancing on one another, have a relatively high rising velocity (0.5–1 m.s^{-1}).

2.9. Several properties of divided solids

2.9.1. *Abrasiveness – classes of abrasiveness*

An abrasive product scratches the surfaces with which it comes into contact, provoking premature wear of the equipment.

The Mohs scale characterizes products, giving them an index rating of between 0 and 10. Index 10 applies to bodies that are scratched by no others: diamonds. If body 1 scratches body 2, it has a Mohs index greater than that of body 2. The term "Mohs", pronounced [*mo:s*], is the name of a German mineralogist. Appendix 1 includes a list of bodies classified according to this index. Of course, we can also define the abrasiveness classes in a more simple way:

1) non-abrasive: Mohs ≤ 3. Plant fats, flours, wood shavings, sawdust, plastic grains, sugar, salt, graphite, de-sanded bituminous coal, gypsum, plant fibers;

2) relatively abrasive: 3 < Mohs < 7. Limestone, de-sanded bauxite, de-sanded lean coal, fertilizer, de-sanded phosphate;

3) highly abrasive: Mohs ≥ 7. Sand, non-de-sanded ore, coke, slag, clinker, mixed ash, broken plant shells, angular gravel, cement, calcinated alumina, concrete.

These abrasiveness classes, like the flowability classes, are only an approximate system.

NOTE.–

Rubber is highly resistant to abrasion (conveyor belts).

2.9.2. Apparent density – friction with wall

Appendix 2 lists the apparent density ρ_a for free and non-compacted, bulk divided solids:

$$\rho_a = (1-\varepsilon)\rho_s$$

ρ_s: the real density (kg.m^{-3});

ε: the porosity of the product.

We observe that flours and powders are denser than the grains from which they are made.

Apparent density establishes the relationship between volume flowrate Q m^3.s^{-1} and mass flowrate W kg.s^{-1}:

$$W = \rho_a Q.$$

A product conveyed in a chain conveyor presents resistance R to displacement due to its weight:

$$R = \rho_a g \Omega tg\phi_p$$

Ω: the volume of solid present in the device (m^3);

g: the acceleration due to gravity (m.s^{-2});

R: the resistance N;

ϕ_p : the angle of friction with wall.

This resistance is a source of energy dissipation. In screws, the power related to the presence of a product is proportional to its apparent density.

2.9.3. Relationship between equipment and product flowability

1) Elevator buckets or rotary vane sectors must be able to empty easily. Consequently, the product should have high flowability.

2) The product should be able to flow unimpeded through chutes, airslides and vibrating conveyors. It should have high flowability.

3) The product has to fill the troughs of elastomer conveyor belts and not escape from flat metal belts. It must have low flowability.

4) Chain conveyors move the product much faster if the product remains as a single block without losing its form too much, which means remaining immobile while the chain crosses it. Products with no flowability are better.

5) Screw conveyors cannot convey products without flowability, as such products would turn in unison with the screw and not move downstream.

2.9.4. Release of dust

Dust can be released on the vertical drop of a divided solid flow.

Plinke *et al.* [PLI 94] studied this phenomenon.

2.10. Storage and transport logistics

Krambrock [KRA 79] studied:

– the appropriate size and type of storage capacities;

– loading lorries or wagons from silos;

– filling and emptying silos;

– measurement of the level of the divided solid in silos.

Thomson [THO 79] conducted a detailed flow study for the various methods of conveyance.

Pneumatic Conveying of Divided Solids

3.1. Pneumatic conveying: introduction

3.1.1. *Description of flow*

When gas velocity (V_G) is high, particles are uniformly spread. This is known as a diluted state.

All divided solids can be transported in their diluted state if the gas velocity is sufficiently high:

1) as V_G decreases, particles tend to aggregate at the lower part of the horizontal tube. On the sediment surface, particles move by successive jumps (saltation). When V_G decreases further, wave movement occurs in the pipe by superficial erosion. This is known as the "dune" phenomenon;

2) if V_G decreases further, the wave height (dunes) increases, reaching the top part of the tube, where it forms blockages or dense phase plugs are separated by pockets of gas. This is known as the blockage regime (or plugging). This situation, accompanied by vibration, can be favored if the product is cohesive;

3) at a very low V_G, there are no more gas pockets between the plugs. This is known as the mobile-bed regime.

We will state here that the mobile-bed regime, in which the divided solid is in its gathered state (also called dense), is preferable. In this situation:

– we reduce the power used to displace the gas and the solid;

– if the divided solid is hard (Mohs hardness; see Appendix 1), we reduce tube wear;

– the production of fines is reduced.

3.1.2. Compacting of a sediment (or de-aeration of a fluidized bed)

Let us split the porous divided solid bed into n horizontal sections and number them from 1 to n from top to bottom. *All of these sections contain the same quantity of solid.*

We will express the moisture of an index q section by the equivalent thickness of fluid contained in it. Let e_q be this thickness.

All sections are assumed to have the same surface area of 1 m^2.

The parameter e_s denotes the volume of solid in section q.

Section q is subject to weight P_{pq} of $q - 1$ sections located above this section:

$$P_{pq} = (q-1)p_s$$

Resistance to compaction increases with the decrease in the quantity of liquid. For the sake of simplicity, we accept a linear development for this value and write the proportionality relationship between P_{rq} and the difference $(e_{qo} - e_q)$ that *increases with the decrease in e_q*:

$$P_{rq} = k\, e_s \left(e_{qo} - e_q\right)$$

Here, coefficient k acts like a spring *stiffness*, which reacts to compression at this difference, to which compression is irreversible.

This stiffness can be compared to the slope at the origin of the divided solid's compression curve when dry (see section 1.1.2, in [DUR 16a]).

The physical significance of ke_s is close to that of the coefficient K of Fargette *et al.* [FAR 97]. For these authors, this is the inverse of σ_c/σ_1 of the flowability coefficient k_E presented in section 1.3.7, in [DUR 16d].

The fluid flow crossing section q does not change its moisture. On the contrary, section q's flattening is expressed by de_q/dt, and is slowed by viscous flow proportional to ΔP_{vqp}. Therefore, according to Darcy's law:

$$\frac{\Delta P_{Vq}}{e_s} = -\frac{\mu}{2K}\frac{de_q}{dt} > 0$$

μ: fluid viscosity (Pa.s);

K: permeability (m^2).

Finally, the behavior of moisture e_q is shown by the equilibrium:

$$P_{pq} = P_{rq} + \Delta P_{Vq} \tag{3.1}$$

At equilibrium, ΔP_{Vq} is zero, and:

$$(q-1)p_s = k\,e_s\left(e_{qo} - e_q^*\right) \tag{3.2}$$

or rather:

$$e_q^* = e_{qo} - \frac{(q-1)p_s}{ke_s} = e_{qo} - \frac{(q-1)g\rho_s}{ke_s}$$

We observe that at equilibrium and according to our hypotheses, moisture decreases linearly from moisture e_{q0} at the surface to moisture close to zero at the sediment base.

The whole equation of dynamic equilibrium is written as:

$$(q-1)p_s = k\,e_s\left(e_{qo} - e_q\right) - \frac{e_s\mu}{2K}\frac{de_q}{dt} \tag{3.3}$$

Let us subtract equation [3.2] from equation [3.3] member by member and simplify by e_s:

$$0 = k\left(e_q^* - e_q\right) - \frac{\mu}{2K}\frac{d\,e_q}{dt}$$

That is:

$$\left(e_q - e_q^*\right) = -\frac{\mu}{2Kk}\frac{de_q}{dt}$$

On integrating:

$$e_q = e_q^* + \left(e_{q0} - e_q^*\right)\exp\left(-\frac{2k\,K}{\mu}t\right)$$

Sediment height corresponds to q = n. Its total height is:

$$H_T(t) = ne_s + \sum_{q=1}^{n} e_q(t)$$

Final equilibrium is given by:

$$H_T^*(t) = ne_s + \sum_{1}^{n} e_q^* = H_s + H_T^*$$

Finally:

$$H_T(t) = H_T^* + \left(H_{TO} - H_T^*\right)\exp\left(-\frac{2k\,K}{\mu}t\right) \qquad [3.4]$$

Experience confirms this exponential evolution.

3.1.3. *Measurement of stiffness, k*

Values μ, ρ_s and K can be obtained independently. In particular, permeability K is expressed by:

$$K = \frac{\varepsilon^3 d_p^2}{150(1-\varepsilon)^2}$$

d_p: the harmonic mean of particle diameters (m);

ε: the fluid volume taken to the total volume.

The value retained for K will be that of the sediment, which is supposed homogeneous, and whose thickness is the arithmetic mean of its thicknesses at the beginning and end of compacting.

The quantity under the exponential of equation [3.4] is written as:

$$\frac{2k\,K}{\mu}t = \frac{t}{t_a}$$

Compacting time t_a is measured experimentally on semi-logarithmic paper. From this, we deduce the stiffness value k as:

$$k = \frac{\mu}{2\,K t_a}$$

3.1.4. *Various regimes of horizontal pneumatic conveying*

Several authors [SAN 03, FAR 97] together with Mainwaring and Reed [MAI 87] have reached the following conclusions:

– only three types of pneumatic flow exist according to the state of the divided solid:

  - diluted state,

  - semi-dense state (plugging),

  - dense continuous state (mobile bed);

(In the semi-dense regime, dense phase plugs move, while the plugs are separated by pockets of gas.)

– There seems to be a general consensus that only two types of measurement suffice to characterize flow type:

   - de-aeration time,

   - permeability (or the percolation coefficient).

By definition, the de-aeration time (t_{da}) is, according to equation [3.4]:

$$t_{da} = \frac{\mu}{2k\,K} \qquad\qquad [3.5]$$

K: permeability divided solid of the (m^2);

μ: fluid viscosity (Pa.s);

k: stiffness of the divided solid deposited (s^{-2}).

So, let us define the percolation coefficient as follows:

$$C = \frac{K}{\mu}$$

C: the percolation coefficient ($m^2.Pa^{-1}.s^{-1} = m^3.s.kg^{-1}$).

De-aeration time t_{da} is deduced according to equation [3.5] as:

$$t_{da} = \frac{1}{2Ck} = \frac{1}{2Ck_o\rho_s} \qquad\qquad [3.6]$$

Chambers *et al.* [CHA 98] showed that if t_{da} is high, then the divided solid can be transported in the mobile bed phase. If t_{da} is low, however, plugging will occur. This confirms the work of Mainwaring and Reed [MAI 87].

Equation [3.6] shows that t_{da} is inversely proportional to product Ck. Percolation C *inversely* varies from t_{da} and, in order to find criteria that vary significantly, we must use the N_C ratio of both values C and t_{da}:

$$N_C = \frac{C\rho_s}{t_{da}} = k_o \times 2(C\rho_s)^2 = \frac{1}{2k_o t_{da}^2}$$

Note that this ratio is adimensional.

Chambers *et al.* [CHA 98] showed that the possible flow regime is deduced from the following data:

$N_C < 10^{-4}$ mobile bed

$10^{-4} < N_C < 10^{-2}$ diluted state

$N_C > 10^{-2}$ plugging

Supposing that constant k_o only depends on the form and size of the solid particles, we observe that the de-aeration time t_{da} is the value determining the flow regime.

In order to interpret these results, we will retain $N_C = 2\, k_o\, (C\rho_S)^2$:

– if $N_C < 10^{-4}$, permeability is low, hence low gas flow occurs through the particles. Stiffness k_o is low and the divided solid bed is supple. Thus, it retains its integrity. Hence, it is a mobile bed;

– if N_C is just above 10^{-4}, flow is higher and the stiffness remains low. The bed dilates up to the diluted homogeneous phase;

– if N_C is just below 10^{-2}, stiffness is significant and the divided solid fragments into isolated parts that are transported in diluted phase;

– if $N_C > 10^{-2}$, percolation C and stiffness are high, which entails a loss of elasticity in the solid bed, which breaks apart, resulting in plugging.

NOTE.–

We must verify by the DEM (discrete element method, see section 4.1 [DUR 16d]) the concepts of:

– flowability (see sections 2.3.8 and 5.1.3 in [DUR 16a]);

– stiffness (see sections 3.1.2 and 2.1.2 in [DUR 16a])

We will not provide this here.

In addition, we can re-read:

– the interaction between two particles (see section 6.3 in [DUR 16b]);

– deformation of a divided solid under compression (see section 5.2.8 in [DUR 16a]).

Of course, two influential parameters must be taken into account:

– roughness of particle surface;

– particle shape assessed by non-sphericity.

Recall that non-sphericity is the relationship of the particle surface to that of a sphere of the same volume.

Both parameters explain the tendency of two particles to stick together.

NOTE.–

Figure 2.6 in [DUR 16b] is applicable here, given that pneumatic conveying can occur in either the dense phase or diluted phase. A decantation flowrate density can correspond to two values of the solid volume fraction in the slurry.

3.2. Properties of divided solids

3.2.1. *Density and porosity*

If ρ_S is the real density of particle material, the apparent density is:

$$\rho_a = \rho_S (1 - \varepsilon)$$

where ε is the empty fraction (porosity) of the product. We can distinguish porosity after compacting by vibration or repeated impacts ε_{ot} (which is, for ε, an absolute minimum) from the aerated porosity obtained by a shower of the product from a height in the order of 50 cm. Maximum aerated porosity is:

$$\varepsilon_{ot} < \varepsilon < \varepsilon_{oa}$$

$$\rho_{oa} < \rho_a < \rho_{at}$$

Porosity on starting fluidization is connected to the angularity level of the particles:

$$\varepsilon_{fc} = \frac{0.415}{\phi_s^{1/3}} = 0.415\Phi_{np}$$

where Φ_s is the sphericity ratio, the inverse of the non-sphericity ratio (Φ_{ns}). For aerated porosity, we have approximately:

$$0.85 \, \varepsilon_{fc} < \varepsilon_{oa} < 0.90 \, \varepsilon_{fc}$$

Porosity depends not only on the particles' shape (angularity), but also on their size.

For fines ($d_p < 100 \, \mu m$) $0.5 < \varepsilon_{oa} < 0.65$

For grains ($d_p > 500 \, \mu m$) $0.4 < \varepsilon_{oa} < 0.45$

Fines (and particularly ultrafines with $d_p < 10 \, \mu m$) have a highly aerated structure that subsists without collapsing because the forces of mutual attraction between the particles are more influential than weight.

A divided solid's "compressibility" is characterized by the Hausner index (I_H):

$$I_H = \frac{\rho_{ot}}{\rho_{oa}} = \frac{1-\varepsilon_{ot}}{1-\varepsilon_{oa}}$$

The I_H of ultrafine products can exceed 1.25 and these products are termed highly compressible as their highly aerated structure can collapse under the effect of impacts or vibrations.

3.2.2. *Geldart classification*

This concerns the behavior of divided solids crossed by a gas current, which Geldart distributes into four groups:

1) Group C (C for cohesion):

These are ultrafine products, whose mean size is <10 µm. The interparticle attractions dominate relative to gravity. During a fluidization operation, aggregates can form easily, with gas having preferential passage between these aggregates. Pneumatic conveying is difficult at reduced velocities, i.e. in dense phase, as aggregates appear and tend to deposit, thereby leading to blockages. In order to destabilize aggregates, there must be sufficient velocity (diluted phase conveying). However, aggregates are less stable if the solid matter's real density is high, indicating that it is not unrealistic to wish to transport cement in dense phase.

2) Group A:

These products are traditionally used in fluidization as a catalyst in the catalytic cracking of petrol, for instance, where $d_p = 60$ µm.

These products are not fine enough to develop any undesired cohesion, but are sufficiently fine to have low permeability so that at the end of a fluidization process, they present a durable gas retention (the bed slowly sags). Moreover, if we increase the gas flow, a group A fluidized bed dilates significantly before bubbles appear. Because of its low permeability, a highly dilated dense phase can coexist besides bubbles.

Group A is easy to convey pneumatically in dense phase (catalytic cracking or 20 µm fly ash).

3) Group B:

$$\rho_s^{0.934} d_p^{0.8} > 1 \quad \text{and} \quad \rho_s d_p^{1.24} < 0.23$$

Particles here are larger than those in group A. Gas retention (holdup) is ephemeral, with bubbles quickly appearing in a fluidized bed without any prior expansion of the bed. These products are easy to convey in dense phase, although it is preferable to transport sand, which is in group B, in the diluted phase.

4) Group D:

This corresponds approximately to what we would call grains:

$$\rho_s d_p^{1.24} > 0.23$$

So, for a plastic material:

$$\rho_s \#1,000 \text{ kg.m}^{-3} \qquad d_p = 2.10^{-3} \text{ m}$$

$$1,000 \times \left(2.10^{-3}\right)^{1.24} = 0.45 > 0.23$$

Group D particles "plug" easily in dense phase conveyance, which is used for plastic grains. The velocity range is not much higher than that corresponding to the apparition of bubbles or gas pockets in a tube. The bubble or pocket apparition rate for group A products is greatly surpassed, with "fluidization" occurring here, or more precisely "fast" or "turbulent" conveyance (although still remaining in dense phase). A phase inversion occurs and the gas phase becomes continuous, surrounding the aggregates with ("packets" of) particles.

3.2.3. *Precautions and choice according to products*

3.2.3.1. *Cohesion products*

These are nearly ultrafine products (particle diameter < 25 μm) such as carbon black, titanium oxide or, more generally, like the great majority of pigments and colorants. These products have low flowability as they are largely consolidated (compressed) and, furthermore, they are difficult to fluidize as they provoke the development of water conduits.

These products are attracted by the lining under the effect of electrostatic or Van der Waals forces. They can accumulate there, which reduces the pipe cross-section and increases the pressure drop. Accordingly, use of the conveyance in dense phase is advised in order to significantly decrease (and almost avoid) friction that leads to static electricity. Expedition occurs from a pressurized tank.

The expedition tank must have a grid with a large surface area, corresponding to a significant airflow relative to the mass of the product present, which is consequently highly diluted. The grid is attached to the cylindrical part of the reservoir. Dispatching occurs from below via an axial cylindrical downcomer.

This is how we convey cement, fly ash, cereal flour, barite, bentonite and pigments in dense phase. The first two of these products are abrasive.

3.2.3.2. Products with low cohesion (flowing) (relatively fine)

These products are composed of particles whose diameter is between 60 and 150 μm. The negative effects of static electricity (sparks, deposit on lining) are more limited here and conveyance is possible at high velocities in diluted phase. The product is taken via a plunging cane into the pile (cereals) or in a slightly fluidized reservoir. The airflow for fluidization is reduced since the grid surface is attached to the tank's lower cone.

Sand, salt (NaCl), sugar, corn coke, and alumina are among the products conveyed in diluted phase. However, the last two of these products are abrasive and would ideally be transported, as we will see below, in dense phase.

3.2.3.3. Abrasive character

The piping can be coated on the inside with an abrasion-resistant elastomer, while the bends can be coated with corundum cement. We must eliminate feeding systems that include moving parts (screws, rotary valves) in order to avoid premature wear.

As a general rule, the number of bends should be kept to a minimum. In this way, a plunging cane will erode less than extraction from the bottom with a bend.

Pneumatic conveyance in dense phase at reduced velocity is recommended for fly ash, cement, coke and alumina.

3.2.3.4. Product brittleness or softening

Brittle crystals, because of their repeated impact with the lining, produce a significant number of fines. On the other hand, some plastic grains head on slipping and on friction with the lining can warm and soften, producing streaking material (angel hair).

A low velocity is required to avoid such problems, which necessitates conveyance in dense phase. Screw feeders or rotary vanes should be avoided or considered very carefully. The Venturi tube is preferable.

3.2.3.5. *Combustible nature*

This entails a risk of explosion that must be reduced by the meticulous earthing of the conveyance line. We note that if the relative moisture of a gas is ≥65%, there is no spark risk. However, moisture can have a downside (see below). Finally, as a last resort, we can use nitrogen in a closed circuit as a carrier gas.

3.2.3.6. *Moisture*

If the product is hygroscopic or if it includes a strong proportion of fines susceptible to adsorbing water vapor, the presence of moisture can give the product a sticky nature, making it adhere to the lining, which can cause blockages. It would be preferable to convey a moist product at high velocity, that is, in diluted phase.

3.2.3.7. *Granulometry spread*

Pneumatic conveyance in dense phase is more difficult if the product's size distribution is widely spread, as the optimal velocity ranges are not the same for fines and grains.

3.2.3.8. *Toxic (or hygroscopic) products*

The carrier gas must be air in a closed circuit.

3.2.3.9. *High friction with the lining*

If the slide angle of the product on a plate of material inclined from the horizontal is greater than 45°, there is a danger of sedimentation in the pipe (especially in the bends).

This is the case for products whose fat levels are >10%.

3.3. Operational curves

3.3.1. *Types of operational curves*

3.3.1.1. *Theoretical curves (Figure 3.1)*

Curve (A) corresponds to the flow of gas alone with no solid present. Curves such as (C) represent pneumatic conveying with flow W_S of a *constant* solid along the curve. Conveying in dense phase *sensu stricto* corresponds to the arcs of curves (C) situated to the left of the $\Delta P/L$ minima, that is, to the left of the minima curve (D).

Quite close to curve (D), curve (E) corresponds to irregular operation with frequent blockages. *Curve (F) sets the minimal values for gas velocity in an empty vat V_G*, below which the conveyable solid flow is zero. For each linear pressure drop, $\Delta P/L$ corresponds to a value of this velocity that is the minimum conveying velocity V_{mt}.

Moving from curve (E) to curve (B), the solid progressively fills the tube so that curve (B) corresponds with a tube full of *immobile* solid that is traversed by gas.

In theory, dense phase conveying concerns the area between curves (D) and (E). In reality, this is typically not the case, as designers prefer to avoid the blockage zone. Therefore, *the operational points for the dense phase are often present slightly to the right of curve (D)* but, of course, for high values of flow solid W_S or, which is equivalent for high values of load ratio μ. This explains why in a correlation such as that of Stegmaier [STE 78], $\Delta P/L$ is slightly ascending with V_G and does not descend at all.

3.3.1.2. *Practical representation ([HIL 86a]; Figure 3.2)*

The pressure drop in the line is:

$$\Delta P = P_0 - P_R$$

where P_0 is the original pressure in the conveying line and P_R is the reception pressure.

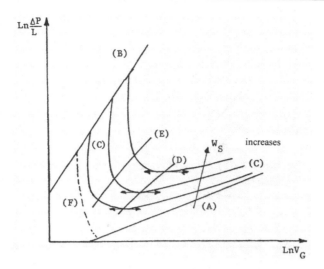

Figure 3.1. *Local pressure gradient according to the local velocity of gas*

In Figure 3.2, the curves correspond to not only a given pressure drop, but also a given pipe diameter (this is the same as fixing tube length).

V_{G0} is the gas velocity on entry to the line.

We observe that the curves $W_S = f(V_{G0})$ present a maximum W_{SMax} for a certain value of V_{G0} and, if $W_S < W_{SMax}$, there are two possible values of V_{G0}. The highest of these corresponds to the diluted phase and the lower to the dense phase.

NOTE.–

Correspondence of these two representations can be reached as follows:

– by cutting the first (3.1) with a horizontal $\Delta P/L = $ Cste (or $P_0 = $ este). In this way, we obtain the second representation;

– by cutting the second (3.2) with a vertical $W_S = $ Cste. In this way, we obtain the first representation.

In Figure 3.2, the operational area often used for the dense phase is usually located not below the W_{SMax} curve (as should theoretically be the case), but slightly above it.

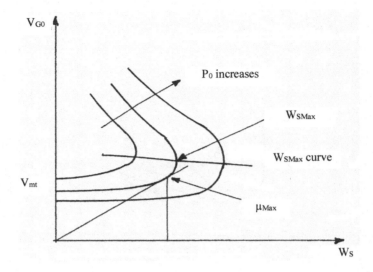

Figure 3.2. *Conveying velocity on line entry according to solid flow*

The pressure drop in the line is:

$$\Delta P = P_0 - P_R$$

where P_0 is the pressure at the origin of the line and P_R is the reception pressure.

In Figure 3.2, the curves correspond to not only a given drop in pressure, but also a given pipe diameter (this is the same as fixing tube length).

V_{G0} is the gas velocity on entry to the line.

We observe that the curves $W_S = f(V_{G0})$ present a maximum W_{SMax} for a certain value of V_{G0} and, if $W_S < W_{SMax}$, there are two possible values of V_{G0}. The highest of these corresponds to the diluted phase and the lower to the dense phase.

In Figure 3.2, the operational area often used for the dense phase is usually located not below the W_{SMax} curve (as should theoretically be the case), but slightly above it.

This practice is removed from the theoretical choice that could be envisaged when looking for the W_S maximum (at a given ΔP or P_0) and the maximum load ratio μ obtained by following a tangent from its origin on curve $P_0 = $ Cste. Indeed:

$$\mu = \frac{W_S}{W_G} = \frac{W_S}{\rho_{G0}Q_{G0}} = \frac{W_S}{V_{G0}}\left[\frac{RT_0}{A_TP_0M}\right]$$

A_T: the cross-section of the tube (m²).

NOTE.–

Chambers *et al.* [CHA 98] proposed (conclusion, page 318) expressions for the minimal velocity of V_{mt} conveying (which they call v_1) according to the conveying regime of the divided solid. They considered velocity v_1 to be the minimum velocity for particle entrainment (dragging along).

3.3.2. *Power expression according to Hilgraf [HIL 86a, HIL 86b, HIL 88] in dense state*

Hilgraf uses what are known as conveyance curves (Figure 3.3). On the horizontal axis, he applies the mass flowrate W_S of the divided solid conveyed and on the vertical axis, the velocity of gas on tube entry. Each curve corresponds to a given value ΔP_R of pressure drop in the line between the compressor and product exit. The variables used are defined as follows:

P_R: the compressor discharge pressure (bar);

P_A: the pressure on line exit (ambient pressure);

Q_{GA}: the gas volume flowrate at ambient pressure;

D_T: the tube diameter;

V_{GA}: the gas velocity on line exit (ambient pressure);

V_{GR}: the gas velocity on discharge from the compressor (line entry);

L_T: the tube length.

We can write:

$$P_R = \Delta P_R + P_A = K_R \frac{W_s^x L_T}{D_T^y} + P_A \qquad [3.7]$$

K_R: the empirical constant characteristic of the product conveyed.

The minimal velocity for conveyance V_{Gmin} is noted by the meeting point of the vertical axis and the extension of the lower arc of conveyance curves. A value of V_{Gmin} corresponds to each value of ΔP_R [HIL 86b].

According to Hilgraf [HIL 86b]:

$$V_{Gmin} = K_V \sqrt{\frac{D_T}{P_R}} \qquad [3.8]$$

K_V: empirical constant characteristic of the divided solid.

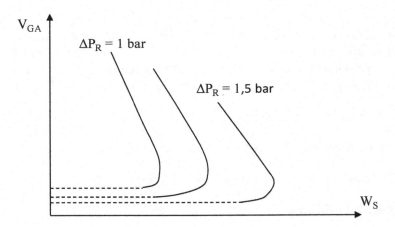

Figure 3.3. *Aspect of conveyance curves*

The gas velocity on entry to line V_{GR} at pressure P_R can be written (for a given ΔP_T) as:

$$V_{GR} = V_{Gmin} + \Delta V_{GR} \qquad [3.9]$$

The pressure drop is denoted by ΔP_R.

The theoretical power for gas compression is:

$$\Pi_{theo} = Q_{GA} P_A \frac{n}{n-1} \left[\left(\frac{P_R}{P_A} \right)^{\frac{n-1}{n}} - 1 \right]$$

where n is the polytrophic exponent, typically taken as 1.3 for air.

We should note here that Hilgraf used the polytropic expression of work, which, by definition, is not isothermal. His equations 2 and 7 essentially show that:

$$Q_{GA} P_A = Q_{GR} P_R \quad \text{or} \quad Q_{GA} = \frac{\pi}{4} D_T^2 V_{GR} \frac{P_R}{P_A} \qquad [3.10]$$

This would imply that compression is isothermal. However, on tube entry, gas comes into contact with the solid, whose thermal capacity is much higher than that of the gas. The gas then cools to a temperature close to ambient temperature and everything happens as if compression had been isothermal (while remaining polytrophic in the compressor).

Using the equations in [3.7], we can eliminate Q_{GA} and P_A. Hilgraf [HIL 86b] found an expression of Π_{theo}, for which he searched for the minimum function of ΔP_R.

In 1988, Hilgraf studied the effect of the presence of successive tube sections, in which the diameter increases from one section to the next to keep gas at a relatively constant velocity in spite of the progressive reduction in pressure.

3.4. Velocity and pressure drops

3.4.1. *Conveying velocity and loading rate – dense and diluted phases*

The loading ratio is the mass flow ratio of the product to the carrier gas:

$$\mu = \frac{W_S}{W_G}$$

Clearly, in economic terms, it is beneficial to maximize the loading ratio.

The conveying velocity is the empty vat velocity of the gas. If the gas expands along the line, we can distinguish the initial conveying velocity V_{t0} and the local conveying velocity V_t.

The minimal conveying velocity V_{mt} is the velocity below which sediment and blockages arise to an unacceptable extent. This velocity depends on the local gas pressure, and is most important for dense phase conveying, which occurs at a reduced velocity and a high loading rate, the opposite of what is used in diluted phase as we will see:

3.4.1.1. *Dense phase conveying*

For this type of conveying:

$$20 < \mu < 100 \qquad\qquad 4\,\text{m.s}^{-1} < V_t < 12\,\text{m.s}^{-1}$$

Dense phase conveying always occurs with air prepared by a compressor. Because of the high loading ratio, the pressure drop is significant (while remaining below 4 bar), and possible line lengths are in the order of a kilometer.

3.4.1.2. *Diluted phase conveying*

Diluted phase conveying can occur in either low pressure (vacuum) or overpressure relative to the atmosphere.

Table 3.1 gives the orders of magnitude.

Operation occurs in vacuum when we are dealing with various origins and a single destination. Operation is recommended under pressure when the line includes one unique origin and several destinations. A mixed operation is possible if both the origins and destinations are multiple. In such cases, the driving engine is placed on the common section.

Some fans can be crossed by the material on condition that the loading rate is <5. Failing this, the fan has to be placed upstream or downstream.

For fibrous products (or fluffy ones like cellulose flakes), the loading rate cannot surpass a value in the order of 10, as these products are very porous ($\varepsilon > 0.98$) and a loading ratio of 10 or 20 corresponds to a simple empty filling of the product at rest, as we will see.

If we accept that the product moves at the same velocity as the gas, we obtain the maximum loading rate, which corresponds precisely to all of the empty spaces being filled by gas as:

$$\mu_{Max} = \frac{\rho_S(1-\varepsilon)}{\rho_G\varepsilon}$$

State of pressure	Type of driving motor	Atmosphere difference (bar rel)	Loading rate	Length of line possible (m)
Depressurization	Vacuum pump	−0.5	$5 < \mu < 15$	500
Depressurization or Overpressurization	Ventilator	−0.1 or + 0.1	$\mu \leq 5$	100
Overpressurization	Lobe compressor	1	$5 < \mu < 15$	500

Table 3.1. *Choice of driving machines*

Spheroidal particles:

$$\rho_S = 2{,}000 \text{ kg.m}^{-3} \quad \varepsilon = 0.40 \quad\quad \rho_G = 3.5 \text{ kg.m}^{-3}$$

$$\mu_{Max} = \frac{2{,}000 \times 0.6}{3.5 \times 0.4} = 570$$

Fibrous product:

$$\rho_S = 1{,}000 \text{ kg.m}^{-3} \quad \varepsilon = 0.99 \rho_G = 1 \text{ kg.m}^{-3}$$

$$\mu_{Max} = \frac{1{,}000 \times 0.01}{1 \times 0.99} = 10$$

We can see why conveyance in diluted phase is always used for fibrous and fluffy products.

3.4.2. *Velocity of air for diluted state*

Muschelknautz and Wojan [MUS 74] proposed an expression relating this velocity to the solid load. However, here we will keep to a very simple empirical relationship that is widely used by manufacturers:

$$V_a = \sqrt{\rho_a}$$

V_a: the air velocity in the tube (m.s^{-1});

ρ_a: the apparent density of the loose product (kg.m^{-3})

$$\rho_a = \rho_s (1 - \varepsilon)$$

ρ_s: the real density of the product (kg.m^{-3});

ε: the porosity of the loose product.

All along the line, pressure drops due to friction and air velocity increases due to decrease in density. In order to retain a constant velocity, certain manufacturers increase the tube diameter in steps.

If the gas is not air, its velocity V_G will be identical to that calculated for air.

Product	ρ_s	ε	ρ_a	V_a
Wood shavings	500	0.6	200	14
Dried sawdust	500	0.8	100	10
Copper turnings	8,800	0.95	440	21
Pulverized coal	800	0.5	400	20
Salt	2,160	0.4	1,296	36
Sand	2,700	0.5	1,350	37
Wheat	1,210	0.4	726	27

Table 3.2. *Product properties (diluted state)*

3.4.3. *Velocity for horizontal conveyance in dense phase*

The practical conveyance velocity is equal to or slightly greater than that corresponding to the maximum W_S for a given pressure drop ΔP.

We suppose that, for $\Delta P \geq 1.5$ bar, the horizontal conveying velocity is given for the products of Geldart's group A as:

$$V_a = V_{th} = 0.24\sqrt{\rho_{oa}} \quad \pm 20\% \ \left(d_p \leq 60\,\mu m\right)$$

ρ_{oa}: the apparent density of the product in aerated state (kg.m^{-3}).

This value must be increased by 50% for products in groups B and D.

We observe that for fine products ($d_p < 60$ μm) belonging to Geldart's group A, the concordance is appropriate.

For type B and D products, the practical experimental velocity is higher than the calculated velocity. This stems from the fact that V_a has to avoid congestion by plugging in the vertical sections, as we will see below.

Product	D (m)	d_p (µm)	ρ_S (kg.m^{-3})	ρ_{oa} (kg.m^{-3})	$V_{a\,exp}$ (m.s^{-1})	$V_{a\,calc}$ (m.s^{-1})
Coal dust	0.085	54	1,500	510	4.8	5.4
Fly ash	"	20	2,380	800	5.4	6.7
Fine limestone	0.085	22	2,730	1,000	7	7.6
Cement	0.40–0.100	25	3,150	1,200	9–11	8.3
Fine crushed quartz	0.100	30	2,600	1,200	9	8.3
Plastic material powder	0.04–0.10	36	950	570	5	5.7
Activated earth	?	45	2,400	800	5	6.8
Sand	0.100	200	2,650	1,350	10	8.8
						$V_{a\,calc.}$ ×1.4
Coarse limestone	0.0825	460	2,730	1,680	12	9.8 13
Plastic material grains	0.262	2,000	950	570	10–12	5.7 8

Table 3.3. *Product properties (dense state)*

3.4.4. *Yang blockage criteria for a vertical tube in dense phase*

We can interpret Yang's [YAN 73] proposal in the following way:

$$\frac{V_R^2}{F+P} = \frac{(V_G - V_S)^2}{F+P} = \frac{V_{IN}^2}{P} \qquad [3.11]$$

with P being the weight of a plug of length L, where:

$$\frac{P}{A_T} = g\rho_S (1-\varepsilon) L$$

F: friction with wall due to solid displacement:

$$\frac{F}{A_T} = \lambda_S \frac{\rho_S (1-\varepsilon) V_S^2}{2} \frac{L}{D}$$

Finally, for the displacement of solid as a cloud:

$$V_{IN} = V_l \varepsilon^n$$

V_l: the limit fall velocity of an isolated particle (m.s^{-1}).

Equation [3.11] is written as:

$$(V_G - V_S)^2 = V_{IN}^2 \left(1 + \frac{F}{P}\right) = V_l^2 \varepsilon^{2n} \left(1 + \frac{\lambda_S V_S^2}{2gD}\right)$$

Yang writes empirically:

$$2n = 4.7 \qquad \text{and} \qquad \lambda_S = f$$

Hence:

$$V_R^2 = (V_G - V_S)^2 = V_l^2 \left[\varepsilon^{4.7}\left(1 + \frac{f V_S^2}{2gD}\right)\right]$$

At the blockage, the relative velocity V_R of the solid relative to the gas is equal to the limit velocity V_l. We have:

$$\frac{f V_S^2}{2gD} = \varepsilon^{-4.7} - 1$$

and

$$V_S^2 = (V_G - V_R)^2 = (V_G - V_l)^2$$

therefore:

$$\frac{2gD\left(\varepsilon^{-4.7}-1\right)}{\left(V_G - V_l\right)^2} = f = 6.81.10^5 \left(\frac{\rho_G}{\rho_S}\right)^{2.2}$$

[3.12]

Otherwise, the solid flow is:

$$W_S = \rho_S \left(1-\varepsilon\right)\left(V_G - V_l\right) A_t$$

[3.13]

Removing $(V_G - V_\ell)$ between equations [3.12] and [3.13], at the jam we obtain:

$$\left(\frac{1}{\varepsilon_E^{4.7}} - 1\right)\left(1-\varepsilon_E\right)^2 = \frac{f}{2gD}\left[\frac{W_S}{\rho_S A_t}\right]^2 = Z$$

[3.14]

Hence, the value of ε_E can be found.

The conveyance velocity to avoid blockages in vertical sections will consequently be obtained by increasing V_{GE} by 40%, so:

$$V_{tv} = 1.4\left[V_l + \frac{W_S}{\rho_S\left(1-\varepsilon_E\right)}\right]$$

NOTE.–

Equation [3.14] leads to a value of ε_E that is clearly less than 1 while we in fact wrote that $V_R = V_\ell$. The gas friction has to compensate for not only the solid's weight, but also friction with the tube lining.

EXAMPLE.–

The gas is air.

In Table 3.4, we see that for the two grains in question:

– where the Froude number $Fr_\ell > 0.123$, engorgement is preceded by plugging;

– conveyance velocity (V_{tv}), see Table 3.4, as we defined, corresponds to the real velocity. On the contrary, velocity V_{th} is too low.

Product	Coarse limestone	Plastic grains
d_p (μm)	460	2,000
ρ_S (kg.m^{-3})	2,730	950
V_ℓ (m.s^{-1})	2.33	5.015
Fr_ℓ	6.7	9.78
D (m)	0.0825	0.262
A_T (m^2)	0.00534	0.0539
W_S (kg.s^{-1})	7.7	73
Z	0.0542	1.26
ε_E	0.905	0.617
V_{tv} (m.s^{-1})	11	12.2
P_{oa} (kg.m^{-3})	1,680	570
V_{th} (m.s^{-1})	13	8
$V_{G\,real}$ (m.s^{-1})	12	11

Table 3.4. *Conveyance velocity*

3.4.5. *Power consumption in a tube*

3.4.5.1. *Solid acceleration*

The solid proceeds a certain distance into the tube, through which it is made to accelerate from zero velocity to a level very close to the velocity of the gas. We will call this the solid acceleration distance L_v.

The gas exerts a drag force F_x on the product situated in length of tube L_v and exerts force dF_x on a part of length dx:

$$dF_x = \frac{\partial F_x}{\partial x} dx$$

So, the force is the product of acceleration γ and mass:

$$\frac{\partial F_x}{\partial x} = \gamma \frac{\partial m_s}{\partial x}$$

In a tube segment with cross-section A and length dx, the mass of solid present is:

$$dm_s = \rho_s R_s A dx$$

R_s: solid retention (fraction volume of occupied by the solid);

ρ_s: the real density of the solid (kg.m^{-3}).

But:

$$R_s A = \frac{W_s}{\rho_s V_s}$$

W_s: mass flow of the solid (kg.s^{-1});

V_s: local velocity of the product (m.s^{-1}).

Removing $R_s A$ between the last two equations, we obtain:

$$\frac{\partial m_s}{\partial x} = \frac{W_s}{V_s}$$

Otherwise:

$$dx = V_s d\tau$$

τ: time (s).

Hence,

$$dF_x = W_s \gamma d\tau, \quad \text{with: } \gamma = \frac{\partial V_s}{\partial \tau}$$

The total drag force is:

$$F_x = \int_0^{L_V} dF_x = W_s \int_0^{L_V} \gamma d\tau = W_s \int_0^{L_V} \frac{\partial V_s}{\partial \tau} d\tau$$

Rather than introducing the solid, V_s is zero, and having run length L_V, the velocity of the solid is practically equal to that of the gas. Therefore:

$$F_x = W_s V_G$$

Now, let us evaluate:

– power lost by the gas;

– power gained by the solid;

– power dissipated in heat.

We will use the fact that power corresponding to a force is the product of this force and the displacement velocity of the point of application of the same force.

Power lost by gas:

$$\Pi_G = F_x V_G = W_s V_G^2$$

Power gained by the solid:

$$\Pi_S = \int_0^{L_V} V_P dF_x = W_s \int_0^{L_V} V_s \gamma d\tau = W_s \int_0^{L_V} V_s \frac{\partial V_s}{\partial \tau} d\tau = \int_0^{L_V} V_s dV_s$$

$$\Pi_s = \frac{W_s V_G^2}{2}$$

Power dissipated in heat is connected to the relative velocity of the gas relative to the solid as:

$$\Pi_C = \int_0^{L_V} (V_G - V_s) dF_x = V_G \int_0^{L_V} dF_x - \int_0^{L_V} V_s dF_x = P_G - P_s$$

$$\Pi_C = \frac{W_s V_G^2}{2}$$

However, only the gas' pressure drop is of interest here:

$$\Delta P_{SV} = \frac{F_x}{A} = \frac{W_s V_G}{A} = \varphi_s V_G$$

φ_s : mass flowrate density of the product ($kg.m^{-2}.s^{-1}$).

As we have seen, the acceleration distance L_V does not enter into these calculations as it is merely an artifice. In fact, strictly speaking, this distance is infinite, but the relative velocity of the gas relative to the solid moves toward zero, so the energy integral is convergent.

3.4.5.2. *Friction of gas on tube*

The Colebrook formula [BRU 68] provides the friction coefficient. For a pneumatic transfer in diluted phase, the regime is clearly always turbulent and the formula is reduced to:

$$f_G = 1/\left(2 \log_{10} \frac{3.71}{\varepsilon / D} \right)^2$$

Irrespective of the material used for the tube, we can safely accept that its absolute roughness is:

$$\varepsilon = 150 \ \mu m = 1.5 \times 10^{-4} m$$

Hence, the pressure drop is:

$$\Delta P_{GP} = f_G \frac{\rho_G V_G^2}{2} \frac{L}{D} \qquad\qquad (Pa)$$

L: the tube length (m);

D: the inside diameter of the tube (m).

3.4.5.3. *Friction of solid on tube*

The Stegmaier formula [STE 78] was refined by Wypich and Arnold [WYP 87] and, accordingly, gives satisfying results (equations 9 and 10 of these authors).

3.4.5.4. *Raising of the product*

Often, the exit of the conveying line is at a higher level than that of the product entry. In other words, the suspension has to be raised to a height ΔH.

If we accept that the solid velocity is very close to that of the gas, the suspension density ρ_D is:

$$\rho_D = \frac{W_S + W_G}{Q_G} \quad \left(kg \ m^{-3}\right)$$

W_S and W_G: the mass flowrates of the solid product and the gas, respectively (kg.s^{-1}).

The power necessary to raise it is equal to the flowrate ($W_S + W_G$) multiplied by the field of the acceleration due to gravity and height ΔH:

$$\Pi_E = \left(W_S + W_G\right)g\Delta H$$

The pressure loss due to the increase in power is:

$$\Delta P_E = \frac{\Pi_E}{Q_G} = \frac{W_S + W_G}{Q_G}g\Delta H$$

Therefore,

$$\Delta P_E = \rho_D g\Delta H$$

3.4.5.5. *Influence of bends*

In the context of pneumatic conveying, bends are designed with a significant curve radius (5–10 times the tube diameter). This precaution has no direct effect on the loss of pressure, but avoids accidental clogging, i.e. clogging by accumulation of the product, reducing the solid flow and with the effect of cancelling it.

On entering a bend, the velocity vector common to the gas and the product changes direction. It changes from $\overrightarrow{V_{G1}}$ to $\overrightarrow{V_{G2}}$. We could suppose that only the gas endures the consequences of this change and that the drop in pressure would be proportional to the vector module ($\overrightarrow{V_{G2}} - \overrightarrow{V_{G1}}) = \overrightarrow{IJ}$.

However, gas effectively has a driving effect in a bend (at the cost of a loss in pressure), while the pipe clearly has a braking effect.

In other words, to go from $\overrightarrow{V_{G1}}$ to $\overrightarrow{V_{G2}}$, the pipe plays the role of absorbing the kinetic energy corresponding to vector $\overrightarrow{FI} = \overrightarrow{V_{G1}} - \overrightarrow{OF}$, while the gas plays the role of creating vector $\overrightarrow{FJ}$. This corresponds to a minimal loss of energy for the gas, as the perpendicular FJ is the shortest distance separating point J from the straight line OI.

The suspension must then acquire velocity:

$$FJ = V_{G1} \sin\alpha = V_{G2} \sin\alpha = V_G \sin\alpha$$

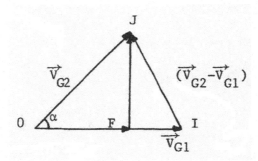

Figure 3.4. *Velocity vector of the suspension*

The description above concerning the acceleration of the solid leads us to write quite simply:

$$\Delta P_{CSV} = \frac{(W_S + W_G)}{A_T} V_G \sin\alpha$$

Half of the corresponding power, that is $Q_G \Delta P_{CSV}/2$, is transferred to the solid in the form of kinetic energy in the new direction, while the other half is dissipated in heat.

In addition, fraction:

$$Q_G = \frac{(W_S + W_G)}{A_T} V_G (1 - \cos\alpha)$$

corresponds to braking on the tube lining and is also transformed into heat.

Finally, the degraded power in the bend is:

$$\Pi_C = Q_G V_G \left(1 + \frac{\sin\alpha}{2} - \cos\alpha\right)\left(\frac{W_A + W_G}{A_T}\right)$$

We have implicitly assumed $\alpha < \pi/2$. If this is no longer the case, we must resolve the change in direction into two steps corresponding to $\pi/2$ and $(\alpha - \pi/2)$. The overall drop in pressure through the bend can be estimated as:

$$\Delta P_C = \left[\frac{W_S + W_G}{A_T}\right] V_G \sin\alpha$$

3.4.5.6. Total pressure drop

Coulson and Richardson [COU 68] in their Figure 7.6 (page 248) give an empirical estimate of the total pressure drop ΔP_T:

ΔP_T	=	ΔP_{SV}	+	ΔP_{GP}	+	ΔP_{SP}
		Acceleration of solid		Friction of gas on lining		Friction of solid on lining
	+	ΔP_E	+	ΔP_C		
		Suspension raising		Bends		

NOTE.– Suspension heating:

The only useful term of ΔP_T corresponds to the suspension raising ΔP_E. The rest corresponds to the product's generation of heat and kinetic energy.

Therefore, the heat that appears is:

$$Q = Q_G \left[\Delta P_{GP} + \Delta P_{SP} + \Delta P_{SV}/2 + \Delta P_C \right]$$

The maximum heating of the suspension is:

$$\delta t = \frac{Q}{W_S C_P + W_G C_G}$$

3.5. Special study of vertical tubes

3.5.1. *Preliminary study*

In his equation 9, Smith [SMI 78] proposed a relationship for a blockage in an n upward co-current flow. Curiously the empty vat velocity of the gas was not used.

Grace and Tuot [GRA 79] studied the initial process of particle cluster emergence. This emergence plays a significant role in some pneumatic conveying and in fast fluid beds.

Leung *et al.* [LEU 71] proposed a linear relationship between velocities V_S and V_G for a blockage in an upward co-current flow. However, this relationship is particularly imprecise.

3.5.2. *Numerical simulation of Desgupta et al. [DES 94a, DES 94b, DES 98]*

These authors use:

– two equations for the conservation of matter;

– two equations implicating quantities of movement (momentum).

To this, we must add the relationships of closure between certain parameters that characterize turbulent flow.

This leads to the following results:

1) the concentration of particles is highest near the lining (segregation) and the dissipation of energy due to gas friction on the particles is much lower than the viscous dissipation within the gas itself;

2) in an upward co-current flow, the authors again find the re-circulation conditions and even flow blockage;

3) they establish the relationship $f(V_S, V_G) = 0$. Clogging for counter-current flows (their Figures 4 and 5 from 1998);

4) in an initially counter-current regime, if we increase the downward solid flow, the gas flow changes direction and occurs downward.

3.5.3. *Minimal length of solid plugs (simplified estimate)*

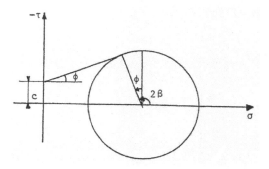

Figure 3.5. *Mohr's circle*

Let us consider a product whose large rupture strain (LRS) can be represented with Coulomb's simplified analytical form:

$$\tau = c + \sigma \tan\phi$$

According to the Mohr system of coordinates (Figure 3.5), we see that:

$$2\beta = \frac{\pi}{2} + \phi$$

β: the angle of principle stress with the normal of the plane of rupture (slip plane);

ϕ: angle of internal friction for rupture of static equilibrium (limit angle of static friction).

We accept that the main major stress is *vertical*.

By rotating $\pm\dfrac{\pi}{2}$, we observe that angle β is also the angle of the horizontal with the slip plane.

Figure 3.6(a) shows that a significant part of the plug can slip along the plane of rupture without hitting the lining, which leads to the end of the blockage, whereas the blockage depicted in Figure 3.6(b) is stable.

The minimum height ensuring stability is consequently:

$$L_{Bm} = D\ tg\beta = D\ tg\left(\frac{\pi}{4}+\frac{\phi}{2}\right)$$

D: tube diameter.

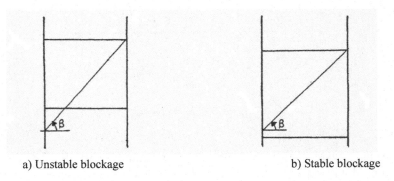

a) Unstable blockage   b) Stable blockage

Figure 3.6. *Blockage stability*

The angles of internal friction seldom exceed 40° for fragmented divided solids such as those used in the chemical industry. A higher limit of L_{Bm}/D would be:

$$\frac{L_{Bm}}{D} \leq tg\left(45° + \frac{40°}{2}\right) = tg65° = 2.14$$

This result, which was probably disregarded by Davidson *et al.* [DAV 85, p. 225] at that time, corresponds to their hypothesis, according to which $L_{Bm}/D = 2$.

3.5.4. *Crossing solid plugs with gas*

We defined the empty vat velocity V_T required to cross solid plugs with gas. The Ergun formula (see Chapter 2 in [DUR 16b]) allows us to express the drop in gas pressure per unit of length of blockage:

$$\frac{\Delta P_B}{L_B} = a\ V_T + bV_T^2$$

or:

$$a = \frac{150\eta(1-\varepsilon)^2}{d_p^2 \varepsilon^3} \quad \text{and} \quad b = \frac{1.75\rho_G(1-\varepsilon)}{d_p \varepsilon^3}$$

η: the gas viscosity (Pa.s);

d_p: the mean harmonic diameter of particles (m);

ρ_G: the gas density (kg.m^{-3});

ε: the porosity of solid plug, with

$$1.1\varepsilon_o < \varepsilon < 1.2\varepsilon_o$$

ε_0: the porosity of the loose solid.

3.5.5. *Friction of solid plugs on a pipe*

1) In divided solid mechanics, we see that, on the pipe, the ratio of the horizontal compression stress to the vertical compression stress is:

$$K = K_a = \frac{\sigma_h}{\sigma_v} = \frac{1 + \sin\delta\ \cos2\beta_a}{1 - \sin\delta\ \cos2\beta_a} < 1$$

This expression assumes that the state of plug stress is of the active type, that is, in principal stress:

$$\sigma_h < \sigma_v$$

We also assume that friction on the wall is entirely involved in the friction coefficient, $\text{tg}\,\phi_p$.

In these conditions (see section 1.4.4 in [DUR 16a]),

$$2\beta_a = \pi - \Delta + \phi_p \quad \text{with} \quad \sin\Delta = \frac{\sin\phi_p}{\sin\delta}$$

δ: the angle of internal dynamic friction.

2) Let us consider the interface separating a solid plug from the gas pocket on top of it. The absolute velocity V_I of this interface is greater than that of the absolute velocity V_{SP} of particles in the pocket. Consequently, the interface "receives" the mass flowrate:

$$\rho_S\beta(V_I - V_{SP})$$

But the same interface "abandons" this same flow in the blockage, since the blockage particles' absolute velocity V_{SB} is less than V_I. Therefore:

$$W_{ST} = A_T\rho_S(1-\varepsilon)(V_I - V_{SB}) = A_T\rho_S\beta(V_I - V_{SP}) \qquad [3.15]$$

When particles go from the pocket to the plug, their velocity increases by:

$$(V_{SB} - V_{SP}) > 0$$

The frontal compression stress σ_F exerted by the accelerated particles on the blockage is the product of the increase in velocity and mass flow density:

$$\sigma_F = \frac{W_{ST}}{A_T}(V_{SB} - V_{SP}) = \rho_S(1-\varepsilon)(V_I - V_{SB})(V_{SB} - V_{SP}) \qquad [3.16]$$

From equation [3.15], we take:

$$V_I - V_{SB} = \frac{\beta(V_{SB} - V_{SP})}{1 - \varepsilon - \beta}$$

Substituting this into [3.16], we have:

$$\sigma_F = \frac{\beta\sigma_S(1-\varepsilon)(V_{SB} - V_{SP})^2}{1 - \beta - \varepsilon}$$

3) Therefore, A_T is the tube cross-section at elementary height dL:

– the solid weight is:

$$\gamma A_T dL$$

– the effort variation for vertical compression is:

$$A_T d\sigma_v$$

– a braking stress τ at perimeter P acts *downward* on the wall opposite to the ascending movement of the plug. This corresponds to effort:

$$\tau P \, dL$$

The balance of a solid blockage section is written as:

$$A_T d\sigma_v = \gamma A_T dL + \tau P dL$$

So, since:

$$\frac{P}{A_T} = \frac{4}{D}$$

the balance equation becomes:

$$d\sigma_v = \gamma dL + \frac{4}{D}\tau dL$$

But we know that:

$$\tau = \sigma_h \, \tan\phi_p = K\sigma_v \tan\phi_p$$

Hence,

$$d\sigma_v = \left(\gamma + \frac{4K\mathrm{tg}\phi_p}{D}\right)dL$$

On integrating:

$$\mathrm{Ln}\frac{\sigma_v + \dfrac{D\gamma}{4K\mathrm{tg}\phi_p}}{\sigma_F + \dfrac{D\gamma}{4K\mathrm{tg}\phi_P}} = \frac{4K\,\mathrm{tg}\phi_p L_B}{D} = Q$$

and in the plug base:

$$\sigma_v = \sigma_F e^Q + \frac{D\gamma}{4K\mathrm{tg}\phi_p}\left(e^Q - 1\right)$$

the pressure drop of the gas must compensate for stress σ_v (rather than $\Delta\sigma = \sigma - \sigma_F$):

$$\Delta P_B = \sigma_v = \sigma_F e^Q + \frac{D\gamma}{4K\,\mathrm{tg}\phi_p}\left(e^Q - 1\right)$$

The values for these parameters are such that we cannot replace $(e^Q - 1)$ with Q. However, since β is in the order of 0.01–0.03, σ_F is negligible and we write:

$$\Delta P_B = \frac{D\gamma}{4K\,\mathrm{tg}\phi_p}\left(e^Q - 1\right)$$

Furthermore, we can write:

$$\Delta P_B = \Gamma L_B$$

Hence, the equation that provides L_B is:

$$\Gamma = \frac{D\gamma}{4L_B K\,tg\phi_p}\left[e^{\frac{4K\,tg\phi_p L_B}{D}} - 1\right]$$

This equation is resolved by iterations that begin by expressing the member to the right, posing $L_B = L_{Bm}$.

Calculation of V_T, V_I and π:

For this, (see section 3.5.4) we simply write:

$$aV_T + bV_T^2 = \Gamma$$

Hence,

$$V_T = \frac{-a + \sqrt{a^2 + 4b\Gamma}}{2b}$$

and

$$V_I = V_G - V_T + V_S$$

(supposing that $\beta = 0$ in the plugging flow):

$$V_S = \frac{W_S}{A_T \rho_S} \quad ; \quad \pi = \frac{V_S}{(1-\epsilon)V_I}$$

3.5.6. *Practical proceedings*

1) Calculation of L_B:

$$L_B = 2D, \text{ or rather } L_B = D\,tg\left(\frac{\pi}{4} + \frac{\phi}{2}\right).$$

2) Calculation of $\epsilon = 1.2\epsilon_0$.

3) Calculation of Γ.

4) Calculation of V_T, V_I and π.

5) Calculation of ΔP:

$$\Delta P = (\pi \Delta H) \Gamma.$$

Plug height is $\pi \Delta H$, where ΔH is the difference in height.

Gas friction on the pipe is utterly negligible.

EXAMPLE.–

$\phi_p = 30°$ (tg $\phi_p = 0.5773$) $D = 0.262$ m $\epsilon = 0.5$

$\delta = 40°$ $\rho a = 570$ kg.m^{-3} $\eta = 18 \times 10^{-6}$ Pa.s

$V_G = 11$ m.s^{-1} $d_p = 3 \times 10^{-3}$ m $\rho_G = 2$ kg.m^{-3}

$\beta = 0$ $W_S = 73$ kg.s^{-1} $\rho_S = 950$ kg.m^{-3}

$$\Delta = \text{Arc sin} \left(\frac{\sin 30°}{\sin 40°} \right) = 51°$$

$$2\beta_a = 180 - 51 + 30 = 159°$$

$$K = \frac{1 + \sin 40° \cos 159°}{1 - \sin 40° \cos 159°} = 0.25$$

$$L_B = 2 \times 0.262 = 0.524 \text{ m}$$

$$\gamma = 570 \times 9.81 = 5,592 \text{N.m}^{-3}$$

Hence,

$$\Gamma = \frac{0.262 \times 5,592}{4 \times 0.524 \times 0.25 \times 0.5773} \left[e \frac{4 \times 0.25 \times 0.5773 \times 0.524}{0.262} \right]$$

$$\Gamma = 10,623 \text{ Pa.m}^{-1}$$

$$a = \frac{150 \times 18.10^{-6} (1-0.5)^2}{(3.10^{-3})^2 \times 0.5^3} = 600$$

$$b = \frac{1.75 \times 2 \times (1-0.5)}{3.10^{-3} \times 0.5^3} = 4,666$$

Therefore,

$$600\ V_T^2 + 4,666 V_T - 10,623 = 0$$

$$\sqrt{\Delta} = (4,666^2 + 4 \times 600 \times 10,623)^{1/2} = 6,875$$

$$V_T = \frac{-4,666 + 6,875}{2 \times 600} = 1.84 \text{ m.s}^{-1}$$

$$V_S = \frac{73}{950 \times \left(\frac{\pi}{4} \times 0.262^2\right)} = 1.43 \text{ m.s}^{-1}$$

$$V_I = 11 - 1.84 + 1.43 = 10.59 \text{ m.s}^{-1}$$

$$\pi = \frac{1.43}{(1-0.5) \times 10.59} = 0.27$$

With a height difference of 6 m, blockage length is:

$$6 \times 0.27 = 1.62 \text{ m}$$

and the pressure drop is:

$$P = 1.62 \times 10,623$$

$$= 17,209 \text{ Pa}$$

$$P = 0.17 \text{ bar}$$

We see the significant value of the solid's friction with the lining, without which the static pressure would only be:

$$10^{-5} \times 6 \times 570 \times 9.81 \times 0.3052 = 0.1 \text{ bar}$$

NOTE.–

The only flow that we have not examined is the flow arising from gas and solid moving downward with a higher gas pressure in the tube. This flow can be considered trivial.

3.6. Ancillary equipment for pneumatic conveying

3.6.1. Feeding systems for continuous installations

3.6.1.1. Suction pipe

This rod is dipped into the mass to be dispatched. Suction pipes can be rigid, flexible or telescopic. They can be adjusted according to the loading ratio required. Suction pipes are only suited to products lacking cohesion (high flowability), such as wheat and grains in general. This solution is adopted with vacuum installations to take a product from the pile.

3.6.1.2. Venturi

The product is injected into Venturi throat in a depressurized state, which sucks the solid material, thereby facilitating its introduction into the line.

3.6.1.3. Screw

Screws are suited to powdery products and cohesive products, though not abrasive ones (not containing sand). The seal is ensured by the product itself, but we can reduce leaks by using a conical screw that compresses material or rather by fitting a retention valve at the end of the screw that applies in cases of reduced product feed.

A conical screw with decreasing pitch can consume up to 10 times more energy than a traditional screw, that is, almost as much as the air compressor. In any case, the pressure difference between upstream and downstream cannot surpass 3.5 bar, and we can attain a flowrate equal to 250 tonne.h^{-1} of solid.

3.6.1.4. *Injector*

This is a bent tube at the base of a dispatch hopper. Gas is injected horizontally, dragging the solid that descends into the bend. A vertical column of solid with a height of between 1 and 4 m is used to ensure the seal upstream. We can also give the injector the shape of a Venturi that is fed with solid in the neck. The injector is suited to high flowability products (sand, alumina, coal).

3.6.1.5. *Rotary valve with vertical passage*

A horizontal shaft fitted with vanes turns into a cylindrical body. The supply arrives from the upper part and leaves under the valve, aided by the force of gravity.

Each rotary valve is designed for a particular product and flow and it is not recommended the rotational velocity be varied.

For cohesive products, an air leak can help to empty the valve sectors. The leak flow depends on the clearance between the rotor and the valve body, the upstream–downstream pressure difference and product cohesiveness. Sticky products can clog leaks. If the tube is under pressure, the alveoli are filled up to a maximum of 50% since air progresses through the valve counter-current to the product. We also need to allow for a lateral vent for its evacuation. In an installation operating in a vacuum, the alveolus-filling rate can reach 88–90%. The fact that this rate does not reach 100% is simply due to the moving divided solid overrunning. Rotary valves turn at a velocity of 30–45 rev.min^{-1}. This velocity is a decreasing function of the alveolus size. Rotary valves can lead to the formation of fines and present a low tolerance to abrasive products.

Valves that avoid the shearing of soft grains (plastics) exist.

If abrasive fines are present, the bearings have to be protected.

The through flowrate of rotary valves depends on their filling sector (alveolus), which is established according to the upstream–downstream pressure difference, so that a flow of 200 tonne.h^{-1} to a receptor under vacuum is reduced to 30 tonne.h^{-1} if the receptor is pressurized.

Valves are only sealed appropriately if the upstream–downstream pressure difference is <4 bar.

3.6.1.6. *Rotary valve with horizontal passage*

Rather than being perpendicular to the tube as in the previous configuration, the shaft, which is again horizontal, is parallel to the tube (horizontal).

Therefore, the alveoli are easily emptied by the tube's gas flow. The supply occurs, as previously, from above.

This system is suited to products with an inclination to agglomerate (moisture).

3.6.1.7. *Shutter valves*

Two superimposed shutters enclose a chamber. When one shutter is open, the other is closed (they are never open at the same time). The chamber is filled or emptied periodically with each inversion of the shutter position. We can see up to 10 cycles per minute.

These shutters are suited for abrasive or brittle products. Their flow is limited, however, and they soon become cumbersome if we need to exceed several tons per hour.

Ultimately, this is a primitive device that can be suited to waste (wood shavings, ash, oily seed shells, cereal grain husks, wood bark).

3.6.1.8. *Venturi*

This device is associated with pressurized installations. The product is introduced through a Venturi throat by a screw or a vibrating transporter and, at this point, pressure is minimal, being below the ambient pressure even if a machine blows gas through the installation under low pressure. Consequently, the product is not at risk of dispersing to the exterior. The velocity at the throat is between 20 and 30 $m.s^{-1}$.

In a pressurized installation in diluted phase, for which the product supplied must not be discharged by gas, it must be introduced by a Venturi throat. The cross-section of this throat must be such that the static pressure is inferior or equal to the ambient pressure P_A. If V_{GE} and V_{GC} are the gas velocities on entry to the convergent and the neck, the Bernoulli equation is written as:

$$\frac{1}{2}\rho_G V_{GE}^2 + P_G = \frac{1}{2}\rho_G V_{GC}^2 + P_A$$

Let σ be the ratio of the throat cross-section to that of the tube cross-section. We obtain:

$$P_G - P_A = \frac{1}{2}\rho_G V_G^2 \left(\frac{1}{\sigma^2} - 1\right)$$

EXAMPLE.–

$$P_G - P_A = 15,000 \text{ Pa} \qquad \rho_G = 1.3 \text{ kg.m}^{-3}$$

$$V_G = 20 \text{ m.s}^{-1}$$

$$15,000 = \frac{1 \times 20^2}{2}\left(\frac{1}{\sigma^2} - 1\right)$$

$$\sigma = 0.115$$

While accounting for the uncertainties of this calculation, we would select a value of $\sigma < 0.115$ as if we suck in a little air from the outside, which would help in the introduction of the product.

3.6.2. *Expedition by discontinuous loads (dense phase)*

This type of expedition occurs from a pressurized cylindro-conical tank, whose inside bottom is conical. These tanks have the advantage of being able to hold air at 5 or 7 bar, which allows for conveying over longer distances (even for abrasive or friable products) due to the absence of moving parts.

Expedition can occur from the top or the bottom:

– from the top: light and/or powdery products;

– from the bottom: dense products and/or grains.

The drop in pressure connected with expedition is:

– >0.1 bar from the top if the product follows a pipe on the inside of the reservoir (but we can also withdraw it laterally);

– <0.1 bar from the bottom as air enters close to the expedition tube.

– A fluidization grate fitted for top expedition. The pressure drops to 0.05 bar (the grate is a fabric held between two perforated metal plates).

In the solution of expedition from the top with fluidization (Figure 3.7), we must balance the proportions of conveying air and fluidization air. In order to increase the ongoing flow of solids, we must increase the fluidization flowrate.

The two airs mix in the expedition line. We measure pressure P_0 at the start of the line. Typically, the compressor is volumetric, which ensures a constant flowrate of air, and the measured variations in pressure P_0 only express those present in the solid flow. Consequently, if the solid flow is excessive, the upstream pressure P_0 will increase (with the drop in pressure along the whole line), requiring a reduction in the proportion of fluidization air. The maximum capacity is reached when all of the air is directed to the tank.

Inversely, if we reduce the fluidization air too much, we can descend below the minimum fluidization velocity and not send anything at all. If we wish to extend the line, the pressure drop must be higher as well. If $P_0 = P_{G0}$, V_{G0} decreases and a blockage can occur.

In reality, we have three manometers for top expedition:

– one on the top part of the tank;

– one on the general tube for air intake;

– one on the sending tube, 1.5 m away from the tank.

If the size distribution is spread (or rather, if the solid is a mixture of grains and dust), a low flowrate of fluidization air may be required to dispatch the product. In such cases, blockages can occur on the expedition line if it is long. Accordingly, the conveying air has to arrive as close as possible to the tank.

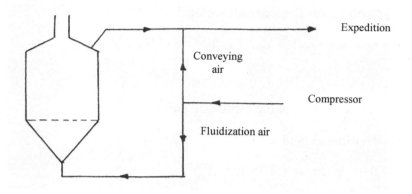

Figure 3.7. *Schematic of top expedition*

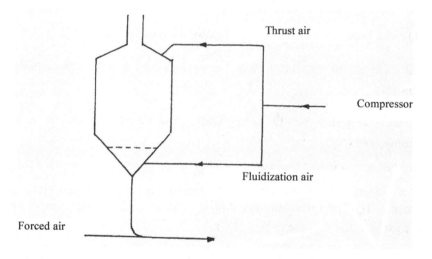

Figure 3.8. *Schematic of bottom expedition*

Bottom sending is suited to products that are difficult to fluidize or to grains that require a very high flow of fluidization air. If the product is cohesive, the compressed air impulse will dislodge any solid blockages.

A single manometer (on the tank) suffices for bottom expedition (Figure 3.8).

3.6.3. *Criteria for the correct design of load conveying*

The total duration of cycle T is the sum of three terms:

$$T = T_R + T_E + T_M$$

T_R: the filling time of the expedition tank;

T_E: the expedition time itself (tank emptying);

T_M: the total dead time.

If Ω_u is the useful reservoir volume, i.e. the volume of a load in the conditions in which it occurs in the tank, we can write:

$$\Omega_u = Q_E T_E = \overline{Q}T$$

Q_E: the instantaneous flowrate of expedition in volume;

$\overline{Q}$: the mean flowrate of solid in volume in the established tank conditions.

Since mean flowrate $\overline{Q}$ is applicable, the investment and exploitation economy leads to:

– a reduction in Ω_u, that is, T;

– a reduction in Q_E (equipment, line diameter, compressors, etc.), which reduces T/T_E. The instantaneous flow Q_E can at times be between 5 and 10 times greater than the mean flow $\overline{Q}$.

Finally, in order to reduce the T/T_E ratio that is still greater than 1, we must reduce dead time T_M and filling time, which is easy and inexpensive to achieve.

3.6.4. *Separation of the product and carrier gas*

1) The simplest solution is a tube discharging onto a pile of the solid, but this system is only usable in pressurized installations and with products that are not dusty.

2) A cyclone:

– either with a sleeve filter;

– or a moist deduster (Venturi, for example).

3) The receptor silo can simply act as a sedimentation chamber (wheat silos), but if the product includes fines, the air exit from the silo must include a dedusting sleeve filter, which must be of modest size.

3.6.5. Conclusion for these systems

We can characterize the upstream–downstream seal of the previous devices by a maximum admissible ΔP.

Broadly speaking, the possible solid supply flow W_S is inversely proportional to the line length L_T:

$$W_s \left(\text{tonne.h}^{-1} \right) = A / L_T \left(\text{km} \right)$$

Supply system	ΔP max (bar)
Suction	– 0.5
Venturi	+ 0.3
Screw	– 0.5 or + 3.5
Rotary valve	– 0.5 or + 4
Shutter valve	+ 2
Pressurized reservoir	+ 5

Table 3.5. Admissible pressure difference

Thus, with a pressurized reservoir and L_T = 200 m = 0.2 km:

$$W_s = \frac{20}{0.2} = 100 \text{ ton. h}^{-1}$$

System	A
Venturi	1
Screw	10
Rotary valve	16
Pressurized reservoir	20

Table 3.6. *Flow coefficient*

3.6.6. *Block valves*

Block valves equip chutes (tube elements that are often close to vertical, in which divided solids can flow via gravity). These valves interrupt either the solid flow alone or that of the gas. We can distinguish between:

1) *gate valves* (with a sliding gate), and in particular guillotine valves, whose shutter is bevel-edged and literally cuts through the solid mass. If particles remain stuck between the shutter and its seat, the valve cannot be sealed to gases, and so it can be useful to follow the valve with another such as a butterfly valve;

2) *butterfly valves* allow blocking of the gas flow. These are economical valves, accounting for the typically large diameters required, and they are direct (almost even integral) passage valves in the open position, which reduces the chances of solid accumulation, thereby hindering closure;

3) *elastomer sleeve valves* or ball valves have the same purpose as butterfly valves, though mostly for limited diameters. Ball valves require non-abrasive products, which is not the case for butterfly valves which can tolerate the passage of abrasive products.

3.6.7. *Tubes for pneumatic conveyance*

If, as is the case in dense phase, there is a significant pressure drop along the line, we must alter the tube diameter and scale it in a series of 2.3 or 4 increasing values.

We must use rolled and welded tubes for non-abrasive products.

The bend radius is often in the order of 5–10 times the diameter of the tube in order to avoid internal erosion in the bend. The straight length following a bend must be reinforced if the product is abrasive.

For abrasive products, the tube must be stretched but not welded, and can be coated on the inside with an elastomer, melted metallic oxide or cement. The Ni-Cr alloy is known for its resistance to abrasion, and it does not require coating.

The switches can be made in flexible tubes.

3.6.8. Receptors (gas–solid receptors)

The receptor silos have a sleeve filter on top that guarantees that the surrounding environment is not contaminated. A cyclone can also be placed between the silo and the filter.

A tangential arrival in the silo benefits the separation of solid and gas by centrifugal effect, but can provoke the erosion of the pipe if the solid particles have a high hardness.

NOTE.–

Some manufacturers propose air injectors at sensitive points all along the line (probable blockage points). Such injections, recommended for conveying in dense phase, must be limited or they transform the dense phase in to a diluted phase. However, these injections can be effective in the case of a blockage.

3.6.9. Control of solid flow: starting and stopping

While it is not generally recommended to act on the rotation velocity of a rotary valve, it is possible to modulate that of a screw or the periodicity of a plug valve.

It is also possible to act on the gas pressure of a pressurized expedition tank. If the pressure drop along the line proves to be less than recommended, by forcing the fluidization air we can increase the solid flow, which increases the value of the pressure drop.

Compressed air enters the installation at a temperature that is generally higher than the ambient temperature, at 60°C for instance. However, when starting, the installation is cold and gas will contract. There is therefore a risk the velocity will fall below the minimal conveying value. A wise precaution would be to begin by circulating the gas by itself in order to establish the installation temperature and to ensure that there is no solid blocking the line.

Inversely, following the end of solid supply, the gas flow must be maintained in order to prevent any accumulation of the product, particularly in the lower parts of the vertical sections. This solid "purge" can take a long time if the product has moderate flowability (cohesive product).

A temporary shutdown of the installation can be provoked by a failure to detect solid in the tube while the expedition valve is open.

3.6.10. *Choice of driving machine for the gas*

The machine can either be a blower (compressor) fed at atmospheric pressure and discharging at a higher pressure, or an aspirator fed at a pressure lower than atmospheric pressure and discharging at atmospheric pressure. Of course, pressurized installations are equipped with compressors and vacuum installations with aspirators.

Fans can be used in the same manner with either compressors or vacuum pumps. The possible overpressurization values communicated to the gas are:

Axial ventilators	$500 \text{ Pa} < \Delta P < 3,000 \text{ Pa}$
Single-stage centrifuge fans	$3,000 \text{ Pa} < \Delta P < 15,000 \text{ Pa}$
Multi-stage centrifuge fans	$15,000 \text{ Pa} < \Delta P < 40,000 \text{ Pa}$

Discharging of cereals from barges and ships can occur with either aspirators or multi-stage fans.

Classic "vacuum pumps" are vane or liquid ring compressors that are used as overpressurizers or aspirators. The liquid ring "pump" can accept a gas that is slightly loaded with dust, and, moreover, a hot dispersion and/or abrasive particles. The vane pump's seal is ensured by lubrication that leads to the installation of an overpressurizer, or if it is an aspirator, the oil must be collected in order to limit its consumption. Vane overpressurizers can discharge up to 4 bar rel.

We require a blower with ΔP not exceeding 0.7 bar, that is, a lobe compressor or centrifuge compressor. These machines, used as aspirators, only tolerate depressurization to 0.4 bar. Centrifuge compressors can be subject to pumping (pulsating operation) just like fans of the same name. This is not the case with lobe compressors. On the contrary, the latter must be protected by an aspiration filter and its flow (on aspiration) cannot exceed 10,000 $m^3 \ h^{-1}$, while centrifugal machines can treat far greater flows.

The screw overpressurizer (dry or lubricated) is, in pneumatic conveying, used at up to 4 bar G to discharge, while the piston compressor can reach 6–7 bar G and can also run dry (graphite or teflon parts) or lubricated.

Hydraulic Conveying of Divided Solids: Standpipe and L-valve

4.1. Power dissipated in slurry

4.1.1. *Introduction*

When treating a clear liquid phase by agitation or pumping, the power required and pressure drop in piping can be precisely calculated.

When we disperse a solid in a liquid phase, the agitation and pumping powers together with the pressure drop are increased due to the friction of the liquid on the solid, friction between solid particles, and friction between particles and walls, stirring blades or pump. In this chapter, we will propose methods for calculating the power and supplementary pressure drops incurred. We will call power or suspension energy the corresponding power and energy. For long-distance conveying of suspensions of ore or coal in water, it is indispensable to know the pressure drop due to the presence of solid.

4.1.2. *Agitation of slurry*

Agitated crystallizers are used to study crystallization by cooling in the laboratory. They are rarely used in industries because of crystal deposition on the cooling coil.

According to Nagata [NAG 75], there are two sorts of agitated recipients:

1) The recipient is equipped with four standard counter-vanes:

The suspension power is:

$$P_s = P_a \left(\frac{\rho_B}{\rho_L} - 1 \right)$$

Therefore,

$$\rho_B = \rho_S c_v + \rho_L \left(1 - c_v \right)$$

c_v: fraction of solid volume;

ρ_B: density of slurry (kg.m^{-3});

ρ_S: real density of slurry (kg.m^{-3});

ρ_L: density of liquid (kg.m^{-3}).

Hence,

$$P_s = c_v \left(\frac{\rho_S}{\rho_L} - 1 \right) P_a$$

Thus, the suspension power is proportional to the volume fraction of solid.

2) The agitated recipient is equipped with four counter-vanes that are situated not on the lateral wall but at the bottom. In this case,

$$P_s = k\, c_v \left(\frac{\rho_S}{\rho_L} - 1 \right) P_a$$

with:

$$k = 0.45 \text{ Ln } d_p + 5.6$$

EXAMPLE.–

$d_p = 150.10^{-6} \text{ m}$ $\qquad$ $\rho_S = 2,000 \text{ kg.m}^{-3}$

$c_v = 0.15$ $\qquad$ $\rho_L = 1,000 \text{ kg.m}^{-3}$

$$k = 0.45 \text{ Ln}\left(0.15.10^{-3}\right) + 5.6 = 1.64$$

$$P_S = 1.64 \times 0.15 \text{ P}_a = 0.25 \text{ P}_a$$

4.1.3. *Pumping power of slurry*

Centrifugal pumps are commonly used for long-distance conveying of slurry. However, in order to ensure circulation in a crystallizer with forced circulation, we use axial pumps with significant flowrate that can reach 10,000 m.h^{-3} (this is the maximum value). In both cases, these are non-volumetric pumps that are capable of attaining the desired height of liquid or, in this case, of slurry.

The shaft power of such pumps is proportional to the product of height and slurry density divided by the yield. The yield is affected by the interaction and friction of solid particles, and is expressed by the coefficient $\eta_c < 1$ given by:

$$1 - \eta_c = c_v\left(2.8 + 0.29 \text{ Lnd}_p\right)$$

c_v: the fraction of solid volume in slurry;

d_p: the mean particle size (m).

This formula expresses the results of the curve network provided by Stepanoff [STE 69]. We see that $\eta_c = 1$ for:

$$d_p = 64.10^{-6} \text{ m}$$

However, there remains the influence of slurry density if d_p is less than this value.

For a single liquid phase, the shaft power corresponding to a given height and flow volume is, as we have seen, proportional to the liquid density:

$$P_a = K \, \rho_L$$

For the same height and volume of flow, the shaft power for slurry is:

$$P_{AB} = K \frac{\rho_B}{\eta_c}$$

The suspension power is the difference between the two:

$$P_s = K \left(\frac{\rho_B}{\eta_c} - \rho_L \right) = P_a \left(\frac{\rho_B}{\rho_L \eta_c} - 1 \right)$$

This formula is acceptable for a draft tube fitted with a marine impeller.

EXAMPLE.–

Consider the case of an axial pump:

$Q = 1,200 \text{ m}^3.\text{h}^{-1}$ $\eta_c = 0.7$ $\rho_S = 2,000 \text{ kg.m}^{-3}$ $c_v = 0.14$

$H = 3 \text{ m CL}$ $d_p = 0.15.10^{-3} \text{ m}$ $\rho_L = 1,000 \text{ kg.m}^{-3}$

$$P_a = \frac{1,200 \times 1,000 \times 9.81 \times 3}{0.7 \times 3,600} = 14,000 \text{ W}$$

$$\eta_c = 1 - 0.14 \left[2.8 + 0.29 \, \text{Ln}\left(0.15.10^{-3}\right) \right]$$

$$\eta_c = 0.965$$

$$\rho_B = 0.14 \times 2,000 + 0.86 \times 1,000$$

$$\rho_B = 1,140 \text{ kg.m}^{-3}$$

Supplement of power imputable to the solid:

$$P_s = 14,000 \left(\frac{1,140}{1,000 \times 0.965} - 1 \right) = 2,540 \text{ W}$$

4.1.4. *Pressure drop in conveying slurry*

We use the Molerys and Wellmann's [MOL 81] method, which is an improved form of the original method proposed by Durand [DUR 53].

4.2. Descent tube

4.2.1. *Standpipe characteristics: use*

A standpipe can be described as follows:

– a vertical tube;

– a divided solid descending through the tube due to the force of gravity;

– there can be absolutely no liquid in the tube;

– the gas pressure at the base of the tube can be greater than that at the top of the tube. The difference between the two is:

$$\Delta P = P_{bottom} - P_{top}$$

The standpipe is mainly used in the following installations:

– evacuating solid at the bottom of a cyclone (cyclone stem);

– descent of solid from a fluidized bed (spout). Here, we can identify two cases:

   - superimposed fluidized beds for drying activated coal counter-current to hot and dry gas,

   - exchange between a fluidized catalytic reactor and the regenerator, which is also fluidized;

– desiling (emptying) a hopper. We will see that the presence of a standpipe can multiply the emptying flow by a factor varying between 2 and 8.

NOTE.–

Practitioners define ΔP as the difference between:

– pressure after tube exit, that is, under the diaphragm where applicable;

– pressure above the product present in the descender's feed hopper.

4.2.2. Description of the divided solid in the tube

The divided solid can be present in three distinctly different states:

– suspension (diluted state), in which porosity is high (>0.8);

– the fluidized state, in which porosity is close (somewhat greater) to that of a fixed bed at rest. We refer to Chapter 6 [DUR 16a];

– the gathered state, in which particles are almost permanently in mutual contact. The collected state can move as a "block". We call this a mobile bed.

For suspensions (diluted state), we must calculate the gas pressure and suspension porosity ε. This also applies to the fluidized state.

For gathered (but not fluidized) states, we must also establish the pressure and the stress states. In a non-fluidized collected state, particles remain in almost permanent mutual contact and porosity does not vary (or varies little).

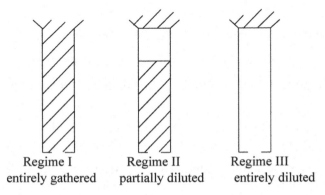

Regime I   Regime II   Regime III
entirely gathered   partially diluted entirely diluted

Figure 4.1. *Various regimes in hatching which represents a collected state and no hatching represents a diluted state*

4.2.3. *Operative mechanism of descenders*

Descenders are typically used with both gases and divided solids, with both directed downward.

It acts, with regard to gas, as a sort of "*hydraulic* joint" that prevents the gas from rising despite the fact that the pressure below is higher than the pressure above. The hydraulic joint effect works as the descending solid drags the gas downward.

The drag force between the gas and the particles is expressed by a weighted mean between the force calculated by the Ergun formula and that calculated by the Richardson and Zaki formula:

$$F = (1 - w) F_{Er} + w F_{RZ}$$

According to Mountziaris and Jackson [MOU 87], we can write:

$$w = \exp\left[\frac{-C_w (1 - \varepsilon)}{(\varepsilon - \varepsilon_o)}\right]$$

For example, when ε tends toward ε_0, w tends toward zero and only the Ergun formula remains.

The weighting coefficient w is arbitrary. We could proceed differently by allowing tortuosity t to apply in the Ergun formula and look for a relationship between t and porosity ε, that is, the Ergun formula can approach that of Richardson and Zaki when ε surpasses 0.5 and tends toward 1 for spheroidal particles. However, we will not do that here.

Now, let us consider the effect of a gas injection.

"Aeration", the injection of gas, is used to ensure that the fraction of gas in the lower part of the tube remains constant. With increasing pressure in this part, the gas density decreases.

Due to the gas descending, the empty vat velocity of the gas phase abruptly goes from value V_G above the level at which it was injected, to value $V_G + V_A$ below the injection level. Consequently, if there is a gathered state above the injection, we will see a diluted state below the injection system.

Now, let us examine why the standpipe placed on a hopper increases the desilage (emptying) flow so significantly.

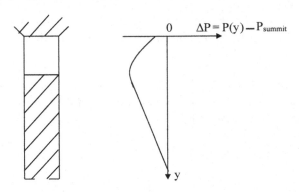

Figure 4.2. *Pressure in regime II of Figure 4.1*

The gas circulates from the top to the bottom, *co-currently* with the solid. According to Yuasa and Kuno [YUA 72], we see that the gas pressure is negative, particularly at the top of the tube, which provokes a suction effect under the hopper (see Figure 4.2) and significantly increases the discharge flow. The negative pressure at the top of the tube is in the order of a tenth of a bar.

This situation only occurs with relatively fine products (e.g. $d_p = 150$ μm) as the divided solid's low permeability causes a strong pressure gradient in the lower part in which the solid is collected, thereby provoking negative pressure in the diluted part.

This phenomenon does not occur with grains ($d = 550$ μm) as the divided solid permeability would be strong and the pressure gradient low. In fact, with grains, regime II does not apply and we have the choice of regimes I and III. An exit orifice, whose diameter is no more than 0.75 of the tube, establishes regime I, and for a tube without diaphragm the only available regime is regime III (see Figure 4.1).

Yuasa and Kuno [YUA 72] proposed their empirical equation 5 to express the flow of solid in a descender.

NOTE.–

Sauer *et al.* [SAU 84] studied the operation of a standpipe used as downcomer between two superposed fluidized beds.

4.2.4. *Slip velocity*

The slip velocity of the solid relative to gas is:

$$V_R = \frac{V_S}{1-\varepsilon} - \frac{V_G}{\varepsilon}$$

V_S: the empty vat velocity of the solid, which is positively directed downward;

V_G: the empty vat velocity of gas, which is positively directed downward.

This velocity can be cancelled. For this, we must have:

$$\frac{V_S}{V_G} = \frac{1-\varepsilon}{\varepsilon}$$

NOTE.–

V_S and V_G are not independent and it is the solid that drags the gas downward.

If the solid flowrate is too weak, then the level of the gas, which is essentially subject to ΔP, will rise in the tube. This is disastrous. The hydraulic joint effect has disappeared.

NOTE.–

Let us distinguish two situations (velocities are positive downward):

1) $\dfrac{V_S}{1-\varepsilon} > \dfrac{V_G}{\varepsilon}$

The solid drags the gas downward, the gas pressure increases and porosity ε decreases.

$$2) \quad \frac{V_S}{1-\varepsilon} < \frac{V_G}{\varepsilon}$$

The solid causes the gas to brake in its downward movement, gas pressure decreases going downward and porosity ε increases going upward. The solid can no longer drag the gas downward.

Of course, velocity V_S is due to the difference in particle weight with the drag force exerted by the gas on the particles.

4.2.5. *Flow-pressure cycles*

1) Let us consider a relatively fine divided solid (154 µm).

When $\Delta P = 0$, the solid flow is significant, but if ΔP increases, the divided solid flowrate will slowly decrease and, at a certain ΔP value, we see an abrupt drop-off in flowrate. Next, if we decrease ΔP, flowrate increases progressively. However, when ΔP reaches a value that we will call minimal, flow abruptly increases and attains a value previously obtained.

Figure 4.3 depicts this cycle (cycle (A)).

When the opening at the base of the tube decreases in diameter, cycle (A) shifts to the *left and downward*.

In the absence of a diaphragm at the base of the standpipe, the divided solid flow is quite simply cancelled if we increase ΔP. This pathway is depicted by line (C) followed by vertical (V).

For the lower branch of cycle (A), flow is always lower than it would be for the discharge from a hopper without a descender. The top branch of cycle (A) corresponds to flows that can reach up to eight times their flow without a descender.

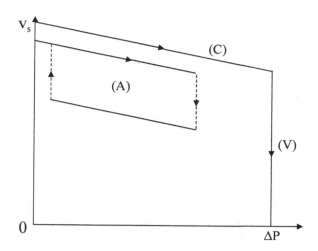

Figure 4.3. *Solid flow; function of ΔP*

2) Now, let us consider a grain (556 μm).

In practice, the top branch of cycle (A) can be obtained by progressively opening the hopper discharge. The lower branch of the cycle can be obtained by progressively opening the tube discharge.

Let us apply the ratio w (reduction) from the diaphragm diameter to the tube diameter.

For a grain, the range of ΔP variation is more limited than that for a fine. The higher slip velocity gives a lower solid retention (higher porosity ε). The increase in discharge flowrate from the hopper is lower than that for a fine (depressurization is less).

For a reduction in the ratio of the lower diaphragm 0.75, there is no diluted–collected interface, and the tube is filled with gathered grains. By progressively opening the discharge, we follow the lower branch of the performance curve in Figure 4.4. The bed is gathered along this curve.

With a reduction ratio of 1 for the exit diaphragm, coexistence can occur between the two divided solid states with the gathered state at the bottom.

NOTE.–

We should note that if the feeding control corresponds to a weak flow of solid, the gas will change direction and move upward, since it is not dragged enough by the solid. This is disastrous. It can lead to the solid causing a blockage or to arching (or doming).

4.2.6. *Principle of performance curve calculation*

The ΔP difference and the diaphragm diameter are the two operational parameters of standpipes.

The drag force exerted by the gas on particles is supposed to act according to the Richardson and Zaki law (see section 4.1.5 [DUR 16b]) or according to a law of weighted means between this law and that of Ergun (see section 1.1.2 of [DUR 16b]).

The system is supposed to be isothermal and the gas ideal. The dependent parameters are:

– porosity and pressure for the diluted state;

– porosity and compression stress (the divided solid is assumed to be cohesionless) for the gathered state.

The principle of calculation is to write the first-order differential equations of the four previous variables (equations 11–14 of Chen *et al.* [CHE 84]).

Resolution of these equations gives ΔP by establishing a value for the solid flowrate.

The performance flow curve fixes solid flowrate according to the following equation:

$$\Delta P = P_{base} - P_{peak}.$$

Pressure P at the bottom is called the counter-pressure.

A family of performance curves characterizes standpipe operations, in which each curve corresponds to a diaphragm opening at the bottom of the tube. This curve family is the performance diagram.

To obtain relatively precise elements in order to perform complete calculation for an installation, we can refer to Chen *et al.* [CHE 84].

Rangachari and Jackson [RAN 82] studied flow stability in standpipes based on performance diagrams. They reached a general conclusion that the stable states are such that:

$$\frac{d(\Delta P)}{d\left(\dfrac{V_s}{1-\varepsilon}\right)} < 0$$

When this derivative is cancelled, a shift occurs from one state to another. The lower arc corresponds to the gathered state and the upper arc to the diluted state.

The positive derivatives correspond to unstable (fictional) states.

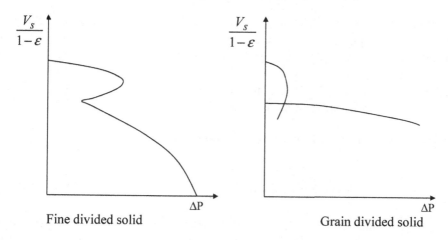

Fine divided solid   Grain divided solid

Figure 4.4. *Performance curves*

NOTE.– Passage from a diluted state to a gathered state

In a diluted state, each particle's slipstream has a greater influence on the following particles as the distance separating them is reduced. A critical distance can be found at a point where a sort of collapse occurs and the particles suddenly cluster.

NOTE.–

Triantafillos *et al.* [TRI 91] used both the calculation and the notion of performance curves to describe the aeration effect. They describe 12 different flow regimes.

4.2.7. *Standpipe simulation by calculation*

Two groups of researchers, Chen *et al.* [CHE 84] and Triantafillos *et al.* [TRI 91], studied the solid flow variations both theoretically and experimentally, showing that the qualitative correspondence between the two methodologies is good and even that the qualitative correspondence was frequently excellent.

The integration of continuity equations and movement equations allows us to place the interface between two types of flow (gathered and diluted). More precisely, the medium's porosity ε (volume fraction occupied by gas) varies very quickly. The slope of function $\varepsilon(y)$ can be described as "steep". We refer to section 2.9 [DUR 16c] for the mathematical resolution of the problem.

We can define coefficient μ of standpipe influence as:

$$\mu = \frac{\text{flowrate with standpipe}}{\text{flowrate without standpipe}}$$

We see that for a given value of ΔP, there are two values of μ: one less than 1 and one higher than 1. *We should not assume that the type of flow in the system is determined according to the value of μ relative to 1.*

In reality, the flowrates in the diluted part and the gathered part in regime 2 are the same.

NOTE.–

In the diluted state, the suspension is not homogeneous and the solid particles can descend in "tongues", which are long clusters whose fall velocity is higher than that of isolated particles. This undermines the quantitative previsions.

NOTE.–

When the opening of the exit diaphragm decreases, the cycles shift vertically downward.

4.2.8. Aeration studies

Shioji et al. [SHI 92] studied solid flow variations according to aeration conditions.

Triantafillos et al. [TRI 91] showed that if we alter the aeration flow, the solid flow goes by a maximum. Flow can be multiplied by 10, but this requires significant aeration.

Takeshita et al. [TAK 92] studied the effect of aeration done close to the standpipe entry.

4.2.9. Placing gas injection at the top of the tube

Let us assume that the gas is injected at the top of the tube.

Takeshita et al. [TAK 92] demonstrated that the solid flow decreases linearly with the injection of gas flow:

$$W = W_o \left(1 - \frac{Q}{Q_{max}} \right)$$

Q_{max}: the aeration flow such that solid flow W_s is cancelled ($m^3.s^{-1}$);

Q: the flow injected ($m^3.s^{-1}$).

In their equation 11, the authors gave the value of Q_{max}.

Now, supposing that the gas is injected into the lower part of the tube, let us denote the pressure above the tube exit as P_o and the ambient pressure as P_a. We can write:

$$\Delta P_o = P_o - P_a$$

Introducing the gas just above the tube exit will destroy any arch or dome that could have formed. Thus, the particles are accelerated and their flow remains stable and regular.

Their equation 8 is written as:

$$W_s = W_{so} + KD_T^{2.5} g^{0.5} \rho_a \left(\frac{\Delta P_o}{D_T \rho_a g} \right)^n$$

The authors provided the values of K and n for particles with a diameter <150 μ.

The injection effect is less significant for grains.

Knowlton and Suhtla [KNO 81] studied the effects of injecting at an intermediate height. However, they worked with a constant solid flow and used gas flows upward and downward. Strictly, this is not a standpipe study, as in a real standpipe, gases move downward.

4.2.10. *Injection study: theory and experience*

Gas injections spread along the tube compensate for the reduction in the empty fraction that occurs under the effect of pressure when we near the lower end of a tube.

However, Triantafillos *et al.* [TRI 91] only studied one single injection at the mid-height of the tube. The objective was to maintain flow in the dense state by preventing plugging. However, *immediately beneath the injection, there is an area where the state is diluted.*

We will not go into the exhaustive details provided by the authors, limiting ourselves here to a few general points.

If the diluted particles descend faster than the gas, the pressure of the gas increases downward and inversely.

In a mobile bed, the pressure development is monotonous. On the contrary, the axial stress can go by a maximum.

The criterion for the formation of a mobile bed from the diluted state is that the empty fraction must diminish sufficiently.

With a tube full of solid gathered in a mobile bed, if we increase the orifice diameter at the bottom of the tube, a diluted state may appear just below the hopper. The fast particles compress the gas and the mobile bed appears below.

However, the theoretical forecast is not rigorous, since the solid flow in a diluted state is very heterogenous, which indicates that the increase in pressure is diminished relative to a uniform spread. This corresponds to a fall limit velocity higher than that for an individual particle.

4.2.11. *Conclusion: standpipes in industrial processes*

Let us simply cite:

1) hydrocarbon synthesis from CO and H_2: the Kellogg process;

2) coal gasification and liquefaction: the Sasol process;

3) catalytic cracking of heavy hydrocarbons: the Houdry and Fischer–Tropsch processes.

4.3. L-valves

4.3.1. *Description and operation*

We refer to the work carried out by Knowlton and Hirsan [KNO 78].

4.3.2. *Uses*

In general, L-valves are used at the bottom of a standpipe.

However, Bachovchine *et al.* [BAC 81] studied a system where particles are exchanged between two fluidized beds *positioned at the same level*.

APPENDICES

Mohs Scale

Nature of divided solid	Mohs index
Wax	0.02
Graphite	0.5–1
Talc	1
Diatomaceous earth	1–1.5
Asphalt	1.5
Lead	1.5
Gypsum	2
Human nail	2
Organic crystals	2
Soda flakes	2
Slaked lime	2–3
Sulfur	2
Salt	2
Tin	2
Zinc	2
Anthracite	2.2
Silver	2.5
Borax	2.5
Kaolin	2.5
Litharge	2.5
Bicarbonate of soda	2.5
Copper (coins)	2.5

Slaked lime	2–3
Aluminum	2–3
Quicklime	2–4
Calcite	3
Bauxite	3
Mica	3
Plastic materials	3
Barite	3.3
Brass	3–4
Limestone	3–4
Dolomite	3.5–4
Siderite	3.5–4
Sphalerite	3.5–4
Chalcopyrite	3.5–4
Fluorite	4
Pyrrhotite	4
Iron	4–5
Zinc oxide	4.5
Glass	4.5–6.5
Apatite	5
Carbon black	5
Asbestos	5
Steel	5–8.5
Chromite	5.5
Magnetite	6
Orthoclase	6
Clinker	6
Iron oxide	6
Feldspar	6
Pumice	6
Magnesium (Mg0)	5–6.5
Pyrite	6.5
Titanium oxide	6.5
Quartz	7

Sand	7
Zirconia	7
Beryl	7
Topaz	8
Emery	7–9
Garnet	8.2
Sapphire	9
Corundum	9
Tungsten carbide	9.2
Alumina	9.25
Tantalum carbide	9.3
Titanium carbide	9.4
Silicon carbide	9.4
Boron carbide	9.5
Diamond	10

We can classify the materials according to their hardness as follows:

Soft   1–3

Quite soft   4–6

Hard   7–10

Appendix 2

Apparent Density of Loose Divided Solids (kg.m^{-3})

A1.1. Vegetable products

Nature of product	Grains	Flours
Flax	720	430
Corn	720	640
Cotton	530	400
Soy	700 (shredded)	540
Coffee	670 (green)	450 (roasted, ground)
Wheat	790	
Barley	620	
Rye	720	
Rice	800	
Oat	410	
Clove	770	

A1.2. Inorganic natural products

Nature of product	Grains	Powders (fine ground)
Bauxite	1,360 (mixed)	1,090
Gypsum	1,270	900
Kaolin	1,024 (shredded)	350 (<10 µm)
Lead silicate	3,700	2,950
Quicklime	850	430
Limestone	1,570	1,360
Phosphate	960	800
Wood shavings	350	320 (sawdust)
Sulfur	1,220	800
Iron	4,950 (beads)	2,370 (filings)
Shale	1,390	1,310
Sodium carbonate	1,060	480
Coke	490	430

A1.3. Manufactured products

Sodium bicarbonate powder	690
Borax	1,700
Catalyzer (fluidized petrol cracking)	510
Ash	700
Charcoal (grains)	420
Coal (mixed)	900
Coal (graded)	700–800
Cement (clinker)	1,400
Cement (Portland)	1,520
Copa shavings	510
Copa shavings auger screw	465
Dolomite powder	730
Soap suds	160
Feldspar (crushed)	1,600
Gravel	1,500
Dairy	2,000

Mica (crushed)	210
Bone (crushed)	1,200
Phthalic anhydride flakes	670
Glass beads	1,400
Iron oxide pigment	400
Zinc oxide pigment	320
Shot pellets	6,560
Potato	700
Rubber clippings	370
Sand	1,350–1,500
Salt	1,200
Sugar crystals	830
Copper sulfate crystals	1,200
Superphosphate powder	810

Bibliography

[ARE 88] AREFMANESH A., MICHAELIDES E.E., "Pressure changes at a sudden expansion in gas–solid flow", *Particulate Science and Technology*, vol. 6, no. 3, pp. 333–341, 1988.

[BAC 81] BACHOVCHINE M., MULIK P.R., NEWLAY R.A. *et al.*, "Pulsed transport of bulk solids between adjacent fluidised beds", *Industrial and Engineering Chemistry Process Design Development*, vol. 20, pp. 19–26, 1981.

[BAT 69] BATES L., "Handling of bulk solids by helical screw equipments", *British Chemical Engineering*, vol. 14, pp. 1072–1076, 1969.

[BRU 68] BRUN A., MARTINOT-LAGARDE A., MATHIEU J., *Mécanique des fluides*, Dunod, 1968.

[BUR 67] BURKHARDT G.J., "Effect of pitch, radial clearance, hopper exposure and head on performance of screw feeders", *Transactions of the A.S.A.E.*, vol. 10, pp. 685–690, 1967.

[CAM 85] CAMPBELL C.S., BRENNEN C.E., SABERSKY R.H., "Flow regimes in inclined open channel flows of granular materials", *Powder Technology*, vol. 41, pp. 77–82, 1985.

[CAO 97] CAO Z., "Turbulent bursting-based sediment entrainment function", *Journal of Hydraulic Engineering*, vol. 123, pp. 233–236, 1997.

[CHA 98] CHAMBERS A.J., KEYS S., PAN R., "The influence of material properties on conveying characteristics", *6th International Conférence on Bulk Materials Storage Handling and Transportation*, University of Wollongong, Australia, pp. 309–319, 28–30 September 1998.

[CHE 84] CHEN Y.M., RANGACHARL S., JACKSON R., "Theoretical and experimental investigation of fluid and particle flow in a vertical standpipe", *Industrial and Engineering Chemistry Fundamentals*, vol. 23, pp. 354–370, 1984.

[CLE 73] CLEAVER J.W., YATES B., "Mechanism of detachment of colloid particles from a flat substrate in a turbulent flow", *Journal of Colloid and Interface Science*, vol. 44, no. 3, pp. 464–474, 1973.

[CLE 76] CLEAVER J.W., YATES B., "The effect of re-entrainment on particle deposition", *Chemical Engineering Science*, vol. 11, pp. 147–151, 1976.

[COD 90] C.O.D.E.T.E.C. (CENTRE D'ETUDES ET DE RECHERCHE DES CHARBONNAGES), *Manutention des charbons fins et humides*, Cahier No.13 de l'Utilisation du charbon, Editions Technip, 1990.

[COU 68] COULSON J.M., RICHARDSON J.F., *Chemical Engineering. Unit Operations*, vol. 2, Pergamon Press, 1968.

[CTB 78] C.T.B. (CENTRE TECHNIQUE DU BOIS), *L'aspiration dans les industrues du bois*, Editions C.T.B., 1978.

[DAR 56] DARNELL W.H., MOL E.A.J., "Solids conveying in extruders", *S.P.E. Journal*, vol. 18, no. 4, pp. 20–29, 1956.

[DAV 85] DAVIDSON J.F., CLIFT R., HARRISON D., *Fluidization*, 2nd ed., Academic Press, 1985.

[DAV 87] DAVIS R.H., LEIGHTON D.T., "Shear-induced transport of a particle layer along a porous wall", *Chemical Engineering Science*, vol. 42, no. 2, pp. 275–281, 1987.

[DUR 53] DURAND R., "Ecoulement de mixture en conduites verticales. Influence de la densité des matériaux sur les caractéristiques de refoulement en conduite horizontale", *La Houille Blanche (Grenoble)*, vol. 8, pp. 124–131, 1953.

[DUR 16a] DUROUDIER J.-P., *Divided Solid Mechanics*, ISTE Press, London and Elsevier, Oxford, 2016.

[DUR 16b] DUROUDIER J.-P., *Liquid-Solid Interactions*, ISTE Press, London and Elsevier, Oxford, 2016.

[DUR 16c] DUROUDIER J.-P., *Adsorption – Dryers for Divided Solids*, ISTE Press, London and Elsevier, Oxford, 2016.

[DUR 16d] DUROUDIER J.-P., *Crystallization and Crystallizers*, ISTE Press, London and Elsevier, Oxford, 2016.

[FAR 97] FARGETTE C., JONES M.G., NUSSBAUM G., "Bench scale tests for the assessment of pneumatic conveying behavior of powders", *Powder Handling and Processing*, vol. 9, no. 2, pp. 103–110, 1997.

[FLA 81] FLATT W., "Energie-Einsparung bei pneumatischer Zementförderung mit dem Fluidstat-System", *Zement-Kalk-Gips*, vol. 34, no. 6, pp. 299–301, 1981.

[GIN 80] GINESTRA J.C., RANGACHARI S., JACKSON R., "A one dimensional theory of flow in a vertical standpipe", *Powder Technology*, vol. 27, pp. 69–84, 1980.

[GRA 79] GRACE J.R., TOUT J., "A theory for cluster formation in vertically converged suspensions of intermediate density", *Institution of Chemical Engineers*, vol. 57, pp. 49–54, 1979.

[HIL 86a] HILGRAF P., "Untersuchungen zur pneumatischen Dichtstrom-Förderung", *Chemical Engineering and Processing: Process Intensification*, vol. 20, pp. 30–41, 1986a.

[HIL 86b] HILGRAF P., "Optimale Auslegung pneumatischer Dichtstrom-Förderanlagen unter energetischen und wirtschaftlichen Gesichtspunkten", *Zement-Kalk-Gips*, vol. 39, no. 8, pp. 439–446, 1986b.

[HIL 86c] HILGRAF P., "Untersuchungen zur pneumatischen Dichtstromfördrung über grosse entfernungen", *Chemie Ingenieur Technik*, vol. 58, no. 3, pp. 209–212, 1986c.

[HIL 87] HILGRAF P., "Minimale Fördergasgeschwindigkeiten beim pneumatischen Festofftransport", *Zement-Kalk-Gips*, vol. 40, no. 12, pp. 610–616, 1987.

[HIL 88] HILGRAF P., "Einflussgrössen bei der energetischen Optimierung pneumatischer Dichtstrom-Förderanlagen", *Zement-Kalk-Gips International*, vol. 41, no. 8, pp. 374–380, 1988.

[JOD 83] JODLOWSKI C., "La manutention pneumatique de produits en vrac. Ecoulements diphasiques à faible vitesse", *Industries Alimentaires et Agricoles*, pp. 709–715, 1983.

[JOH 69] JOHANSON J.R., "Feeding", *Chemical Engineering*, vol. 76, pp. 75–83, 1969.

[JOH 78] JOHANSON J.R., "Design for flexibility in storage and reclaim", *Chemical Engineering*, vol. 85, no. 24, pp. 19–26, 1978.

[JON 90] JONES M.G., MILLS D., "Products classification for pneumatic conveying", *Powder Handling and Processing*, vol. 2, pp. 117–122, 1990.

[KAL 74] KALEN B., ZENZ F.A., "Theorical–empirical approach to saltation velocity in cyclone design", *A.I.Ch.E. Symposium Series*, vol. 70, no. 137, pp. 388–396, 1974.

[KLI 87] KLINZING G.E., ROHATGI N.D., ZALTASH A. *et al.*, "Pneumatic transport. A review", *Powder Technology*, vol. 51, pp. 135–149, 1987.

[KNO 78] KNOWLTON T.M., HIRSAN I., "L-valves characterized for solids flow", *Hydrocarbon Processing (International Edition)*, vol. 57, no. 3, pp. 149–156, 1978.

[KNO 81] KNOWLTON T.M., SISHTLA C., "Void-gas stripping in standpipes", *A.I.Ch.E Symposium Series*, vol. 77, no. 205, pp. 184–198, 1981.

[KOR 91a] KORZEN Z., "Zum Fließverhalten feinkörniger Schüttgüter in Trogkettenförderern", *Z.K.G. International*, vol. 44, pp. 182–186, 1991a.

[KOR 91b] KORZEN Z., "Zum Fließverhalten feinkörniger Schüttgüter in Trogkettenförderern", *Z.K.G. International*, vol. 44, pp. 253–256, 1991b.

[KRA 79] KRAMBROCK W., "Lagern und Unschlagen von Schüttgütern in der Chemischen Industrie", *Chemie Ingenieur Technik*, vol. 51, no. 2, pp. 104–112, 1979.

[LEU 71] LEUNG L.S., WILES J.R., NICKLIN D.J., "On the design of vertical pneumatic conveyer systems", *Proceedings of Pneumotransport 1*, Cambridge, 1971.

[LEU 73] LEUNG L.S., WILSON L.A., "Downflow of solids in standpipes", *Powder Technology*, vol. 7, pp. 343–349, 1973.

[LEU 76] LEUNG L.S., "Cocurrent downflow of suspensions in standpiped", in KAIRNS D.L. *et al.* (eds), *Fluidization Technology*, vol. II, Hemisphère Publishing Corp., Washington, DC, p. 125, 1976.

[LEU 78] LEUNG L.S., JONES P.J., "Flow of gas–solid mixtures in standpipes. A review", *Powder Technology*, vol. 20, pp. 145–160, 1978.

[MAI 87] MAINWARING N.J., REED A.R., "Permeability and air retention characteristics of bulk solid materials in relation to modes of dense phase pneumatic conveying", *Bulk Solids Handling*, vol. 7, pp. 415–425, 1987.

[MAR 73] MARSEN J.M., "Flow of fluidized solids and bubbles in standpipes and risers", *Powder Technology*, vol. 7, pp. 93–96, 1973.

[MIL 90] MILLS D., "Keeping pneumatic delivery up to speed", *Chemical Engineering*, vol. 97, pp. 94–105, 1990.

[MOL 81] MOLERUS O., WELLMANN P., "A new concept for the calculation of pressure drop with hydraulic transport of solids in horizontal pipes", *Chemical Engineering Science*, vol. 36, no. 10, pp. 1623–1632, 1981.

[MOU 87] MOUNTZIARIS T.J., JACKSON R., "The effects of aeration on the gravity flow of particulate materials in vertical standpipes", *A.I.Ch.E Symposium Series*, vol. 83, pp. 10–22, 1987.

[MOU 01] MOUNTAIN J.R., MAZUMDER M.K., SIMS R.A. *et al.*, "Triboelectric charging of polymer powders in fluidization and transport processes", *I.E.E.E. Transactions on Industry Applications*, vol. 37, no. 3, pp. 778–784, 2001.

[MUS 74] MUSCHELKNAUTZ E., WOJAHN H., "Conception des systèmes de transport pneumatiques", *Chemie Ingenieur Technik*, vol. 46, pp. 223–235, 1974.

[NAG 75] NAGATA S., *Mixing*, Halsted Press, 1975.

[NIC 62] NICKLIN D.J., "Two-phase bubble flow", *Chemical Engineering Science*, vol. 17, pp. 693–702, 1962.

[NOV 74] NOVAK P., NALLURI C., "Correlation of sediment incipient motion and deposition in pipes and open channels with fixed smooth beds", *"Hydrotransport 3"*. *Third International Conference on the Hydraulic Transport of Solids in Pipes*, Colorado School of Mines, 15–17 May 1974.

[ONE 68] O'NEILL M.E., "A sphere in contact with a plane wall in a slow linear shear flow", *Chemical Engineering Science*, vol. 23, pp. 1293–1298, 1968.

[PLI 94] PLINKE M.A.E., LEITH D., GOODMAN R.G. *et al.*, "Particle separation mechanisms in flow of granular material", *Particle Science and Technology*, vol. 12, pp. 71–87, 1994.

[RAN 82] RANGACHARI S., JACKSON R., "The stability of steady states in a one-dimensional model of standpipe flow", *Powder Technology*, vol. 31, pp. 185–196, 1982.

[RAU 87] RAUTENBACH R., SCHUMACHER W., "Theoretical and experimental analysis of screw feeders", *Bulk Solids Handling*, vol. 7, pp. 675–680, 1987.

[REH 67] REHKUGLER G.E., "Screw conveyors. State of the art", *Transactions of the A.S.A.E.*, vol. 10, pp. 615–621, 1967.

[RIC 54] RICHARDSON J.F., ZAKI W.N., "Sedimentation and fluidisation. Part 1", *Trans. Inst. Chem. Eng.*, vol. 32, pp. 35–53, 1954.

[SAN 03] SANCHEZ L., VASQUEZ N., KLINZING G.E. *et al.*, "Characterization of bulk solids to assess dense phase pneumatic conveying", *Powder Technology*, vol. 138, pp. 93–117, 2003.

[SAU 84] SAUER R.A., CHAN I.H., KNOWLTON T.M., "The effects of system and geometrical parameters on the flow of class-B solids in overflow standpipes", *A.I.Ch.E Symposium Series*, vol. 80, no. 234, pp. 1–23, 1984.

[SHI 92] SHIOJI S., TOKAMI K., YAMAMOTO H. *et al.*, "Control of the bulk flow of granular materials by an aeration technique", *Powder Technology*, vol. 72, pp. 215–221, 1992.

[SMI 78] SMITH T.N., "Limiting volume fractions in vertical pneumatic transport", *Chemical Engineering Science*, vol. 33, pp. 745–749, 1978.

[SOO 82] SOO S.L., CHEN F.F., "The boundary conditions of the diffusion equation", *Powder Technology*, vol. 31, pp. 117–119, 1982.

[STE 69] STEPANOFF, *Gravity Flow of Bulk Solids and Transportation of Solids in Suspensions*, Wiley, 1969.

[STE 78] STEGMAIER W., "Zur beruchnung der horizontaless pneumatischen Förderung feinkörniger Feststoffe", *Fordern und Heben*, vol. 28, pp. 363–366, 1978.

[TAK 76] TAKESHITA T., KANO T., WATANABE H. *et al.*, "An investigation of the falling phenomenon of granular materials, from a vessel. Part I: Effect of the diameter, length and angle of the efflux tube on the flowrate of particles", *Journal of the Research Association of Powder Technology*, vol. 13, pp. 13–18, 1976.

[TAK 92] TAKESHITA T., ATUMI K., VEHIDA S. *et al.*, "Effect of aeration rate on flowrate of granular materials from a hopper attaching a standpipe", *Powder Technology*, vol. 71, pp. 65–69, 1992.

[THO 79] THOMSON F.M., "Smoothing the flow of material through the plant: feeders", *Chemical Engineering (New-York, N.-Y.)*, vol. 85, no. 24, pp. 77–87, 1979.

[TOD 69] TODA M., KONNO H., SAITO S. *et al.*, "Hydraulic conveying of solids through horizontal and vertical pipes", *British Chemical Engineering*, vol. 14, no. 8, p. 1077, 1969.

[TRI 91] TRIANTAFILLOS J., MOUNTZIARIS J., JACKSON R., "The effects of aeration on the gravity flow of particles and gas in vertical standpipes", *Chemical Engineering Science*, vol. 46, no. 2, pp. 381–407, 1991.

[TSU 92] TSUJI Y., TANAKA T., ISHIDA T., "Lagrangian numerical simulation of plug flow of cohesionless particles in a horizontal pipe", *Powder Technology*, vol. 71, pp. 239–250, 1992.

[WEI 91] WEISS D., Modélisation de l'écoulement gravitaire d'un matériau granulaire, PhD Thesis, Institut National Polytechnique de Lorraine, 1991.

[WIR 81] WIRTH K.-E., MOLERUS O., "Bestiwmung des Druckverlustes bei pneumatischer Strähnenförderung", *Chemie Ingenieur Technik*, vol. 53, no. 4, pp. 292–293, 1981.

[WU 03] WU F.-C., CHOU Y.-J., "Rolling and lifting probabilities for sediment entrainment", *Journal of Hydraulic Engineering*, vol. 129, no. 2, pp. 110–119, 2003.

[WYP 85] WYPYCH P.W., ARNOLD P.C., "A standardized-test procedure for pneumatic conveying design", *Bulk Solids Handling*, vol. 5, no. 4, pp. 755–763, 1985.

[WYP 87] WYPYCH P.W., ARNOLD P.C., "Classification and prediction of fly ash handling characteristics for dense-phase and long distance pneumatic transportation", *T.I.Z. Fachberichte*, vol. 111, no. 11, pp. 753–761, 1987.

[YAN 73] YANG W.-C., "Estimating the solid particle velocity in vertical pneumatic conveying lines", *Industrial and Engineering Fundamentals*, vol. 12, no. 3, pp. 349–352, 1973.

[YAN 75] YANG W.-C., "A mathematical definition of choking phenomenon and a mathematical model for predicting choking velocity and choking voidage", *A.I.Ch.E. Journal*, vol. 21, no. 5, pp. 1013–1015, 1975.

[YUA 72] YUASA Y., KUNO H., "Effects of an efflux tube on the rate of flow of glass beads from a hopper", *Powder Technology*, vol. 6, no. 2, pp. 97–102, 1972.

[ZEN 53] ZENZ A., "Visualizing gas–solid dynamics in catalytic processes", *Petroleum Refiner*, vol. 32, no. 7, pp. 123–128, 1953.

[ZEN 64] ZENZ F.A., "Conveyability of materials of mixed particle size", *Industrial & Engineering Chemistry Fundamentals*, vol. 3, p. 65, 1964.

Index

Printed in the United States
By Bookmasters